Langley Laboratory's first wind tunnel, a replica of a 10-year-old British design, became operational in June 1920.

For sale by the U.S. Government Printing Office
Superintendent of Documents, Mail Stop: SSOP, Washington, DC 20402-9328
ISBN 0-16-037924-5

Metal workers welding pipe pause for the camera in this 1929 view.

WINDS OF CHANGE

by James Schultz

EXPANDING THE FRONTIERS OF FLIGHT

Langley Research Center's 75 Years of Accomplishment 1917-1992

National Aeronautics and Space Administration

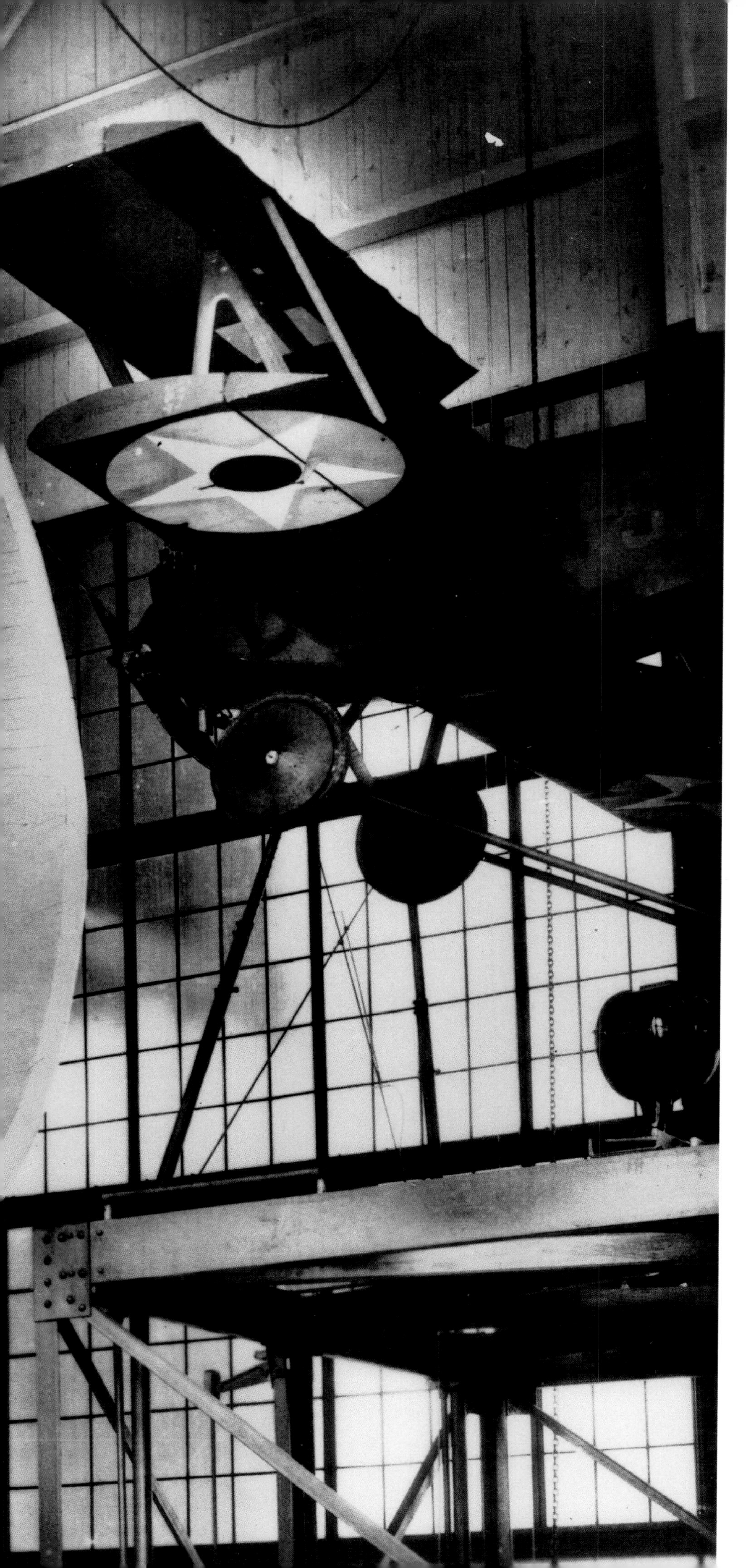

Contents

On the Cover: The image on the cover of this publication was digitally composed from two photographs: one of a National Aero-Space Plane model that was tested at Langley Research Center and the other of the Langley 30- by 60-Foot Tunnel, which has been operational since 1931.

Left: A Langley researcher ponders the future, in mid-1927, of the Sperry M-1 Messenger, the first full-scale airplane tested in the Propeller Research Tunnel.

Introduction

It gives me great pleasure to congratulate the NASA Langley Research Center on its 75th Anniversary in 1992.

This National Research Laboratory occupies a special place in the nation's aviation and space history. Established on July 17, 1917, Langley served as the first research laboratory for NASA's predecessor, the National Advisory Committee for Aeronautics (NACA). In this unique role, the Center bore primary responsibility for nursing U.S. aviation from infancy to world leadership.

With the birth of NASA in 1958, Langley added another unique dimension, this time as a leader in the nation's fledgling Space Program. Langley could look beyond Earth and to the limitless vistas of space. The Center again was on the cutting edge of an exciting new era.

Langley continues to make significant contributions to space and civil aeronautics. The Center's work has made commercial, military and general aviation aircraft safer and more efficient, and has helped make air travel and space exploration a part of every person's life.

On this 75th Anniversary, Langley can look back with pride on its accomplishments and forward to many more. Men and women of Langley Research Center, our nation thanks you. Your pioneering work has been vital to making us first in aviation and space, and I am confident that your continuing efforts will help keep us there.

Congratulations and best wishes…

Daniel S. Goldin
NASA Administrator

A 1951 look into the 16-Foot High-Speed Tunnel.

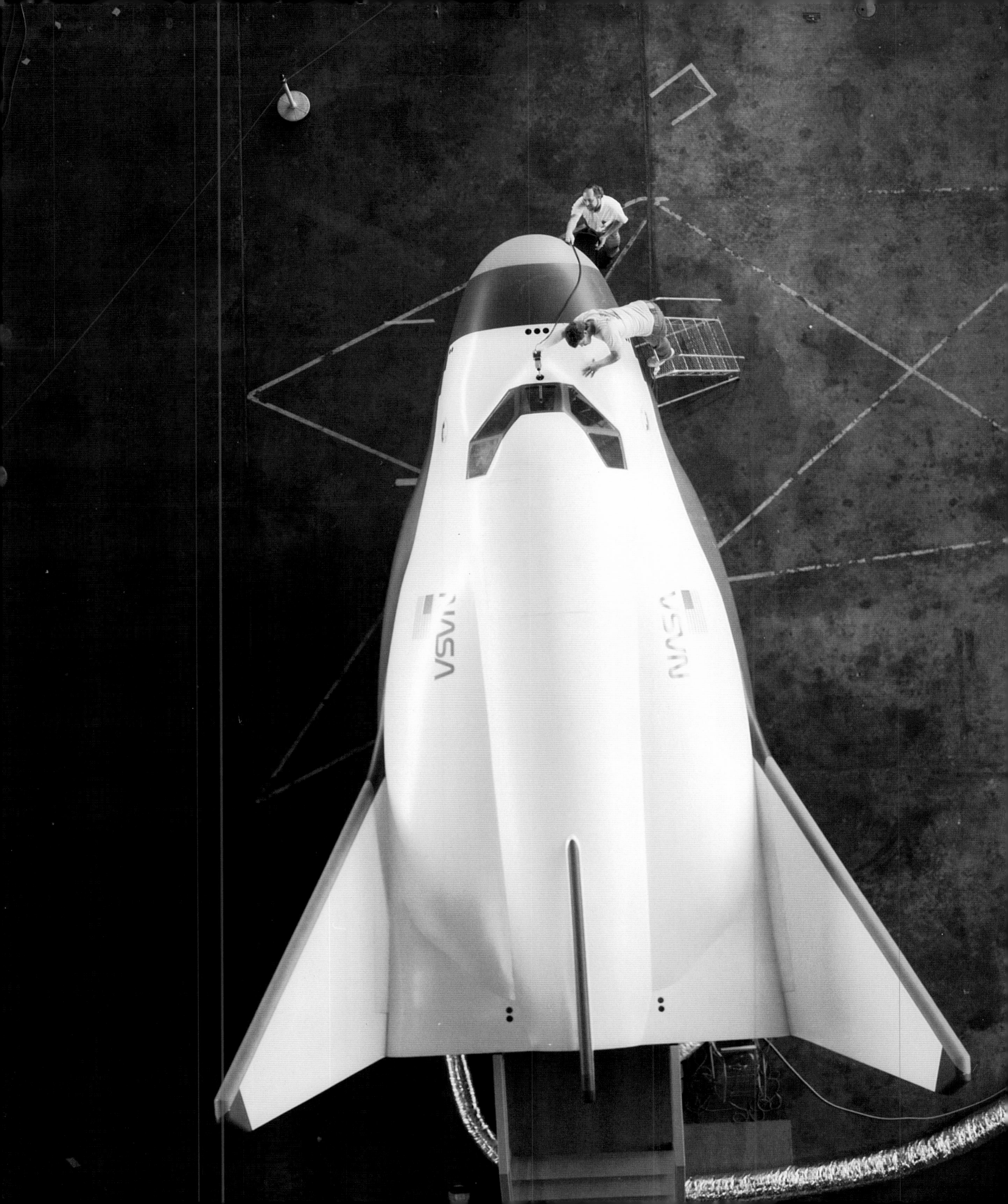
NASA
NASA

Prologue

Seventy-five years ago, in 1917, the Langley Research Center was born. This book is a testament to the years between then and now. In words and pictures, these pages will take you on a trip through time, starting when Langley was little more than pasture and marshland. You will witness the struggles of early aviation, the breaking of the sound barrier and the dawn of the space age. Through it all, you will see the major role that Langley played—and still does. I think you will find it an exciting journey.

The trip begins in 1915 with the creation of the National Advisory Committee for Aeronautics (NACA), NASA's predecessor. Congress established the NACA largely in response to the growing dominance of European aircraft. The Wright brothers had given the United States the early lead with their historic flight in 1903. By World War I, however, it had become clear that the edge had been lost to German, British and French aviation advances.

One of the NACA's first steps toward regaining air superiority was to establish a research center for aviation. Committee members sought a tract of land that was close to the water, not far from the nation's capital, and inexpensive. The search ended in Hampton, Virginia. When Langley Memorial Aeronautical Laboratory, later to become Langley Research Center, was built, it became the nation's first civilian aeronautical laboratory. Langley's mission was forthright: To find practical solutions to the problems of flight.

A lot has happened since then. Who could have imagined that Neil Armstrong and colleagues would practice lunar landings in a cow pasture that once was part of the birthplace of George Wythe, a signer of the Declaration of Independence. Or that, less than a mile away on the historic Chesterville Plantation, Langley scientists in the hypersonic tunnel complex would learn how to break the grasp of gravity and orbit the Earth.

Langley has a proud history and a long list of technological firsts. Langley has hired and trained generations of aeronautical engineers, technicians, managers and leaders, and in the process helped establish the nation's aeronautical infrastructure. From Langley came a group of people who broke technological barriers, created an inventory of aeronautical research tools, helped set up the country's aviation industry, contributed to the establishment of aeronautical departments at universities throughout the nation, and worked to create five of NASA's centers: Ames, Lewis, Dryden, Wallops and Johnson.

Langley's early focus was aviation. But the minds and talents of the Center's work force were soon challenged anew, first by jet propulsion and supersonic flight in the '40s, then by spaceflight in the '50s. Langley achieved major breakthroughs in all areas. Our researchers and test pilots helped break the sound barrier at Edwards Air Force Base. Other Langley researchers were instrumental in designing the Mercury capsule, setting the stage for the Center's leadership role in the space program in 1958. The original seven astronauts trained at Langley. Langley designed the lunar rendezvous technique, proving that two spacecraft could maneuver and dock in orbit. Langley mapped the Moon for the Apollo missions and achieved the only two successful spacecraft landings on Mars.

We can be proud of such a past. We can be equally proud of the present, and we can look to the future knowing that we have the talent, drive and vision to achieve our goals. And what does the future hold? Rest assured that whatever it is, Langley will be in the forefront. Meanwhile, sit back, turn the page, and enjoy your journey through our first 75 years.

Paul F. Holloway
Langley Research Center Director

Paul F. Holloway Langley Research Center Director

Opposite page: Top view of the Langley HL-20 Lifting Body. This concept is a candidate for a future personnel launch system proposed to carry people and small payloads to Space Station Freedom.

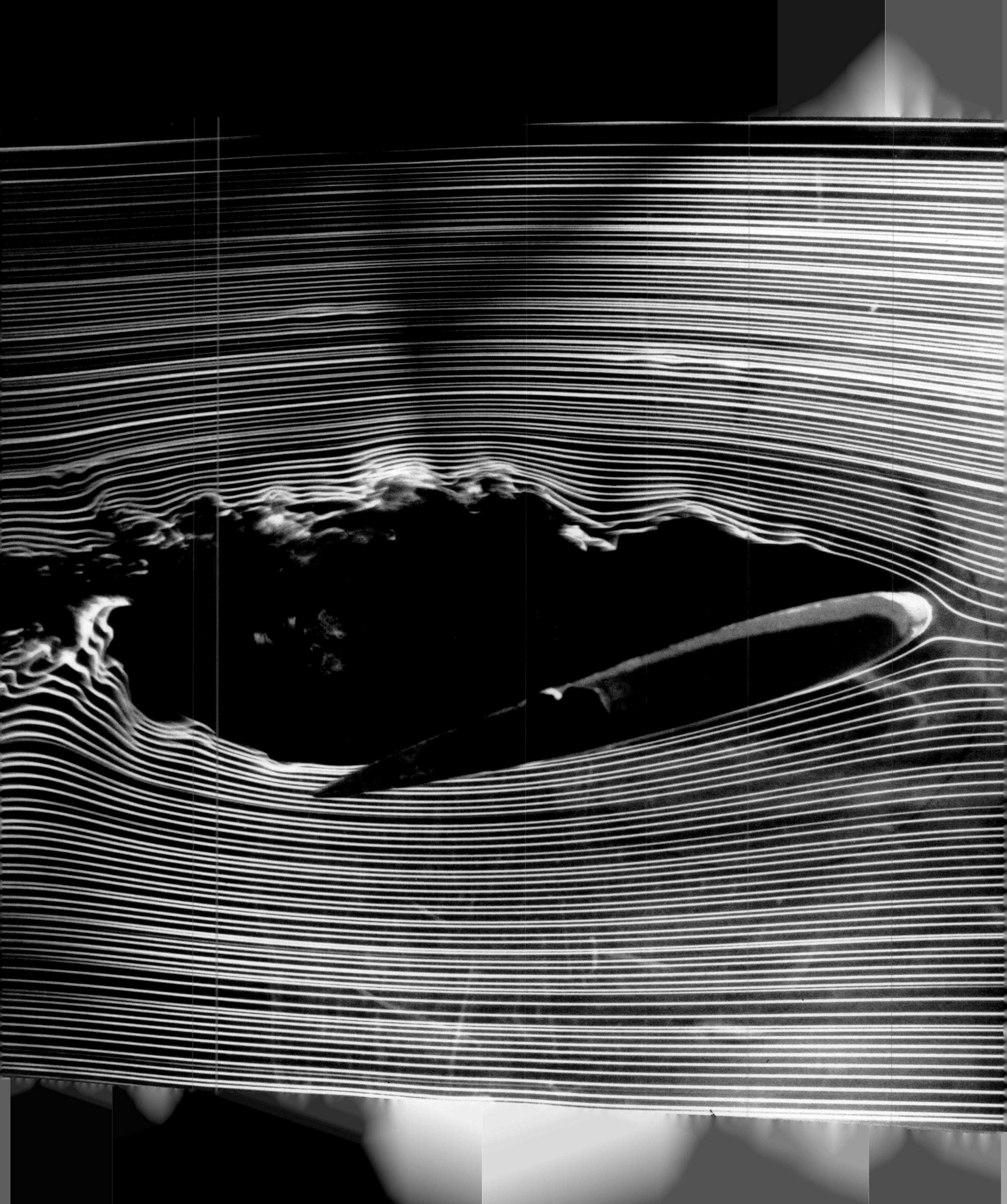

A Laboratory for Flight

When just a young man, novelist Thomas Wolfe set out to see the world. His travels eventually led him to the Virginia port city of Norfolk and, when he heard of work available nearby, to the fishing hamlet of Hampton. There, in the summer of 1918, Wolfe and hundreds of others labored in the oppressive heat and humidity to construct a "Flying Field." In his fictional, semiautobiographical book *Look Homeward, Angel* Wolfe's alter ego Eugene Gant recalls the experience as "the weary and fruitless labor of a nightmare." The workers, wrote Wolfe, reshaped the landscape, blasting ragged stumps from spongy soil, filling the resultant craters that "drank their shovelled toil without end," grading and leveling the ground from dawn to dusk. Meanwhile, overhead, the "bird-men filled the blue Virginia weather with the great drone of the Liberties," practicing aerial observation and photography in British-designed and

Smoke flow visualization shows the flow of air around a model airfoil at 100 feet per second.

Langley Memorial Aeronautical Laboratory as it appeared shortly after completion in 1918.

American-made DeHavilland DH-4s.

All the hard work had a dual purpose: the creation of a new U.S. Army Air Service airfield and the nation's first government-sponsored civilian aeronautical research laboratory. Both were named in honor of Samuel Pierpont Langley, former secretary of the Smithsonian Institution and an avid aeronautical researcher. The research laboratory—Langley Memorial Aeronautical Laboratory—would be overseen by a parent agency, the National Advisory Committee for Aeronautics, or the NACA. The NACA's straightforward mandate was to undertake "the scientific study of the problems of flight with a view toward their practical solution." The new organization would bring together the best of the public and private sectors, creating industry-government partnerships that would in decades to come advance American aviation far beyond its modest beginnings.

NACA's Langley Laboratory would become one of the country's foremost sources for reliable, detailed information on airplane design and performance. Would-be aeronautical engineers attending universities read research papers published by Langley researchers. Both the fledgling commercial aircraft industry and those concerned with the performance of military aircraft looked to Langley for help with all manner of difficulties, from aerodynamic stability and control to structural integrity, from propulsion efficiency to means of reducing drag. As it tackled and solved a variety of problems related to airplanes and flight, all the while paying close attention to detail and displaying a passion for accuracy, Langley established an international reputation as the world's premier aeronautical laboratory.

The Laboratory enlarged its mission in the

late 1950s when the arrival of the Space Age shook the international geopolitical order and promised dramatic new technological possibilities on the "high frontier." A successor agency, the National Aeronautics and Space Administration, or NASA, assumed responsibility for Langley, which was subsequently renamed the Langley Research Center. NASA's mission, like the NACA's, was still geared to aeronautical research, but the new agency's mandate also commanded it to look beyond Earth's atmosphere and to create human-carrying craft that could navigate the unforgiving vacuum of space. "Langley led the way in aeronautical research in the first half of the 20th century," observes current Langley Director Paul F. Holloway, "and in the following decades we would also lead the way in aerospace-related engineering science. In particular, Langley provided NASA with a large part of the engineering and administrative nucleus for the U.S. manned spaceflight program."

"Langley has been responsible in large part for making the United States first in the world in aeronautical technology. From 1920 through 1940 Langley pushed aeronautical technology far beyond where it had been," says former Langley Director Richard H. Petersen (1985–1991), who is currently Associate Administrator for the Office of Aeronautics and Space Technology at NASA Headquarters. "Langley also had a major responsibility in bringing the U.S. into the space era. Project Mercury came out of Langley and much of the Apollo technology came from Langley. Langley people were also involved in the early Space Shuttle conceptual design. Langley was able to assemble together a group of outstanding researchers on the cutting edge of their respective fields and technologies."

Continuing expansion of both the U.S. Army's Langley Field and NACA's Langley Laboratory is evident in this 1933 photograph. The structures in the background with checkerboard roofs are U.S. Army airplane hangars.

Throughout its history and through research and/or applied engineering, the Center has been responsible for some of the 20th century's fundamental aeronautical and aerospace breakthroughs. The nation's first streamlined aircraft-engine cowling was developed at Langley Laboratory. Among other firsts: the "tricycle" landing gear; techniques involving lower-drag-producing flush riveting; development of the sweptback wing; research that aided in breaking the "sound barrier"; origination and design of the Mercury space program; development of rendezvous and docking devices and techniques that made possible the Apollo Moon landing;

The Langley complex as seen in May 1930. Under construction in the foreground is the Full-Scale Tunnel.

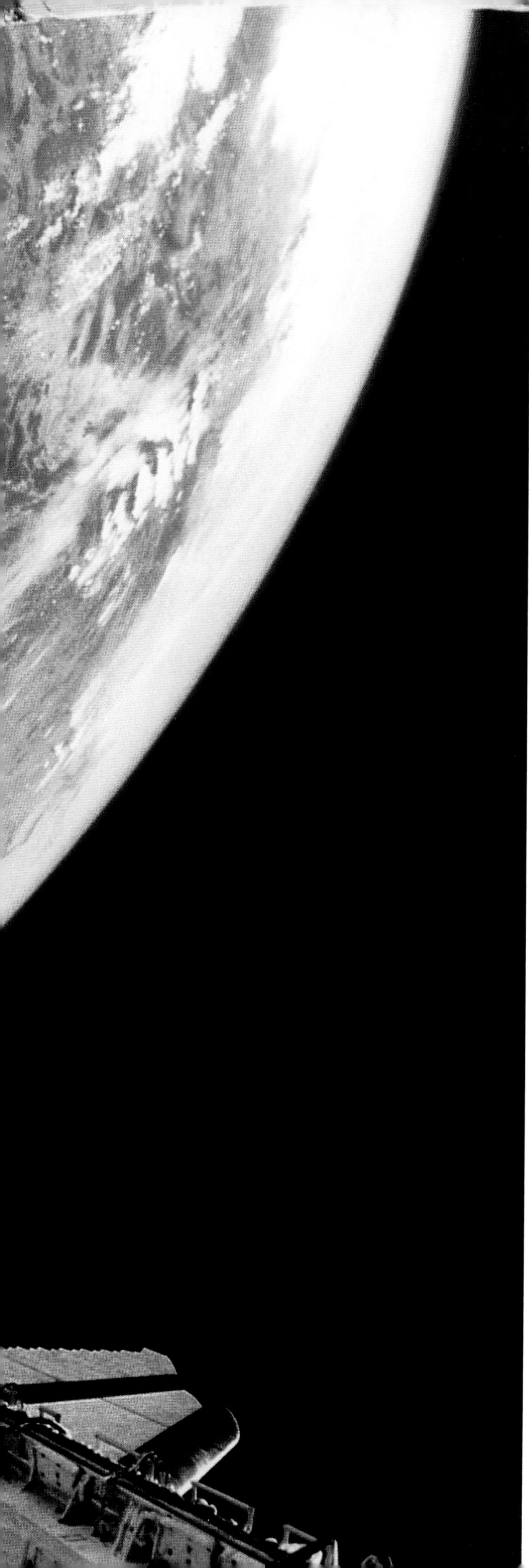

and the design of other unique spacecraft, including a low-cost orbital space-science laboratory, the Long Duration Exposure Facility. In addition, Langley developed and refined instrumentation systems for aircraft, contributed to improvements to aircraft structures and airplane crashworthiness, and, in general, played a major role in the development of generations of advanced military and civil aircraft.

This first chapter is intended to generally acquaint the reader with Langley Research Center; the chapters that follow highlight Langley's technological accomplishments in more detail. Still, the Langley story as told in this volume is abbreviated and incomplete and is offered only as an extended introduction to the nation's oldest aeronautical research facility. It is nonetheless hoped that even this brief jaunt through time will offer perspective on the Center's many achievements in aerospace science and engineering.

At left, the Space Shuttle Challenger *deploys Langley's Long Duration Exposure Facility in this 1984 photo. Mexico's Baja peninsula is visible to the upper left of* Challenger's *open cargo bay.*

Former Langley Research Center Director Richard H. Petersen stands next to a model of the Pathfinder transport, mounted in the National Transonic Facility.

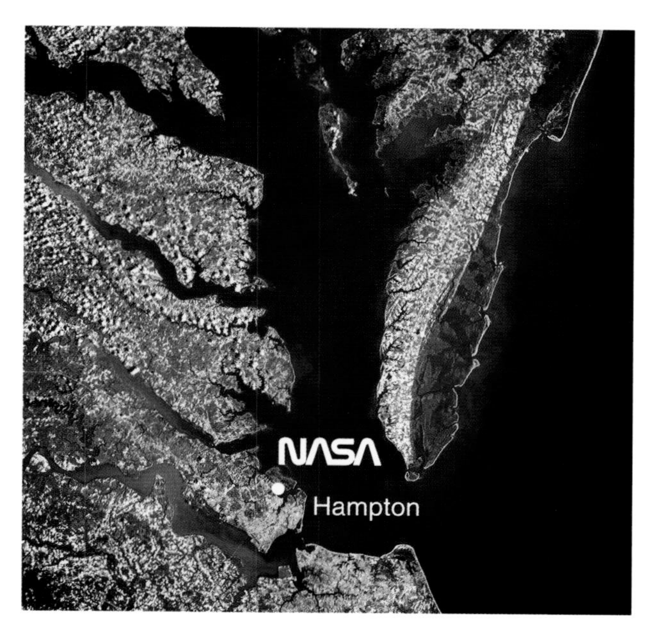

This infrared Landsat satellite view of the Chesapeake Bay watershed includes Hampton, Virginia, home to NASA Langley Research Center.

Taking Flight

Human beings appear to have always been fascinated by flight. Over centuries of recorded history, scores of hardy experimenters tried to take to the air. While some failed less frequently than others, none managed sustained flight in other than unpowered craft—none, that is, until the Wright brothers in 1903. Why did it take so long? One answer might be that, until the latter part of the 19th century, aeronautical researchers spent too much time trying to mimic the wing-flapping flight of birds and not enough time figuring out designs for powered, fixed-wing, human-carrying flying machines. Whatever the reason, achieving human-controlled flight in a powered vehicle was a knotty problem, demanding the expenditure of great technical effort and considerable engineering ingenuity, yet promising little in the way of rewards, tangible or otherwise.

Langley's Lunar Landing Research Facility was used to simulate the Moon's gravity by means of a specially designed crane that supported five-sixths of the weight of a full-scale model of the Apollo Lunar Module. (Designated a National Historic Landmark in 1985.)

One aeronautical pioneer who yearned to become the first to build a full-size, heavier-than-air flying machine was Langley Research Center's namesake, Samuel P. Langley. Already a distinguished scientist by 1886, the year he began his aeronautical explorations in earnest, Langley was fascinated by the challenges flight presented. Despite criticisms from skeptical colleagues, and after 10 years of often frustrating struggle, Langley and his assistants were greatly encouraged by the successful launch of one of their unmanned scale models on May 6, 1896. Vowing to become the first to launch a full-scale, man-carrying airplane, the group intensified its efforts over the succeeding 7 years. However, their final manned test of the Langley Aerodrome on December 8, 1903, ended abruptly in failure, as the awkward machine lumbered not into the air, but into the chill waters of the Potomac River after being catapulted from its original position on top of a houseboat.

The Langley Aerodrome, brainchild of a group led by Samuel Langley. Shortly after this photo was taken, the December 8, 1903, manned tests of the Aerodrome ended abruptly in failure, as it fell into the Potomac River.

Langley Laboratory—the nation's first government-sponsored civilian aeronautical research laboratory—was named in honor of Samuel Pierpont Langley, former secretary of the Smithsonian Institution and an avid aeronautical researcher.

Fated to succeed where Langley failed were Orville and Wilbur Wright, experienced bicycle mechanics and sons of a Midwestern cleric. To prepare themselves for their assault on the record books, the Wrights pored over reports and accounts compiled by aeronautical predecessors and colleagues. Over time the brothers painstakingly pieced together what they believed to be the engineering requirements that must be met if a powered craft was to take off and land under human control. Nine days after Samuel Langley's hopes were doused in the Potomac, on a raw December 17 morning on a desolate North Carolina beach, the Wrights maneuvered their prototype *Flyer* into position for takeoff. The primitive machine flew but a mere 120 feet into the teeth of a stiff wind, but its 12-second flight was enough to propel it into the annals of history.

What the Wrights had wrought was, in a strict sense, unnatural. Walking and running are natural to the human animal; flying is not. But humans had again and again demonstrated an astonishing capability to exceed the limitations of their environment. In designing and flying an airplane, Orville and Wilbur Wright had handily exceeded another human limitation and, in the process, created a potent new symbol for a new century. Moreover, the flight of the *Flyer* was a triumph of engineering science. Not coincidentally, it was applied engineering that would prove to be Langley Laboratory's strength and emphasis.

Although news of the Wrights' achievement was met with disbelief in the several years following their initial flights (the straitlaced brothers believed they should be taken at their word; they limited access to the details of their invention and permitted but a handful of individuals to witness a small number of test flights), by 1908 demonstration flights in France led to worldwide acclaim for the pair. In aeronautical circles it was assumed that the Age of Flight had finally arrived. But in the United States at least, flight continued to be regarded as a sort of goofy indulgence fit for adventurers, daredevils and eccentrics. Even though the first transcontinental flight had taken place by 1911, the prospect of fleets of airplanes carrying paying passengers seemed, to put it mildly, improbable. In the first full decade of the 20th century, Americans of serious temperament dismissed the airplane as a fad or as a specialized machine suitable only for military purpose.

Across the Atlantic, meantime, Europe was well ahead of the United States in aeronautics. In Europe governments were funding ambitious

Frozen in mid-stride by the camera lens, Wilbur Wright watches his airborne brother Orville make history on December 17, 1903.

aeronautical research programs and private firms were designing new generations of airplanes. Americans of vision who were convinced that aviation had a grand future fretted over their country's seeming indifference to federally funded aeronautics research. By 1915, however, the jolt of the first "world" war tumbled the nation out of its aviation-research slumber. On March 3 of that year the 63rd Congress passed a resolution authorizing the creation of a government-sponsored committee to study aeronautics. Thus was born the National Advisory Committee for Aeronautics, which was given $5,000 to begin its aeronautical research.

The NACA was composed of a Main Committee consisting of seven government and five private-sector members. It was intended that the Committee meet in Washington, D.C., semiannually (and more often if necessary) to identify key research problems to be tackled by the agency and to facilitate the exchange of information within the American aeronautical community. The unsalaried Committee, independent of any other government agency, would report directly to the President, who appointed its constituent members. Perhaps too idealistically, it was hoped that members of the Committee would put ego, personal and public

An April 18, 1929, meeting of the Main Committee of the National Advisory Committee for Aeronautics, otherwise known as the NACA. Left to right: J. F. Victory, Secretary; Dr. W. F. Durand; Dr. Orville Wright; Dr. G. K. Burgess; Brig. Gen. W. E. Gillmore; Maj. Gen. J. E. Fechet; Dr. Joseph S. Ames, Chairman; Dr. D. W. Taylor, Vice Chairman; Capt. E. S. Land; Rear Adm. W. A. Moffett; Dr. S. W. Stratton; Dr. G. W. Lewis, Director of Aeronautical Research; Dr. Charles F. Marvin. (One member absent - Dr. Charles G. Abbot.)

agendas, and personality conflicts aside in the interest of advancing aeronautical research. Considering human nature and the inherent limitations of working in committee, the surprising thing was that the NACA Main Committee functioned as well as it did.

"Over the years some members were in effect only honorary, some did not understand research, and some just did not put forth a good effort," writes James Hansen, author of *Engineer In Charge: A History of the Langley Aeronautical Laboratory, 1917–1958*. "On the whole, however, the committee system worked."

As the NACA began its work in Washington, high on the agenda was finding land on which to build its first research laboratory. The Committee's best chance to quickly obtain the required parcel was to cooperate with the Army Air Service, which itself was looking for a site to house an experimental facility with adjacent airfield. The land chosen was 1650 acres just north of the small Virginia town of Hampton. At the time, the site was located in Elizabeth City County, a largely rural area that was home mostly to fishermen and farmers. The land was flat, fronting on water, an advantage when conducting test flights. It was east of the Mississippi and south of the Mason-Dixon line, an area generally prone to good weather and therefore good flying. It was no farther than 12 hours by train from Washington, D.C. Nor was it so close to an unprotected coastal area as to be subject to attack or possible capture in the event of war.

Although the first of the NACA's laboratory buildings was complete by the end of the summer of 1917, the Army's resistance to a permanent civilian aeronautical laboratory (the Army felt the military would do a better job of airplane research than civilians) slowed somewhat the NACA's research timetable. Matters were finally resolved, however, and on June 11, 1920, Langley Memorial Aeronautical Laboratory—and its first wind tunnel, appropriately christened "Wind Tunnel Number One"—was formally dedicated. The NACA was in Hampton to stay.

In a speech delivered before the Air Force Association in Spokane, Washington, on May 31, 1957, NACA Executive Secretary John F. Victory framed the challenges confronting the organization in its early years: "Forty years ago we had just entered World War I and had a great deal to learn. We had but small knowledge of aeronautics—and most of that had come from abroad. We were short of spruce with which we then built planes; short of linen to cover the wings; short on engine power—we had no engine over 80 horsepower. We were short of factories, short of pilots, short of know-how. In short, we were just caught short."

To confront the daunting technological challenges it faced, Langley Laboratory had to build a professional and support staff from the ground up. Early on, the NACA committed itself to finding the best and brightest to solve the problems of flight. Not surprisingly, the caliber of Langley's people would matter most as the Laboratory pushed ahead across the uncharted frontiers of aeronautical research.

A Collective Effort

The young engineers who came to work at Langley in its first decades brought with them a particular sense of mission. Most were aeronautics enthusiasts, interested in all things having wings and propellers. In coming to the Laboratory these aeronautical engineers hadn't chosen a job, but a vocation. Some approached their labors with an almost religious intensity, working nights and weekends with a zeal of which only devotees are capable. The majority kept more regular hours but were no less enamored of the cause. For many, coming to work at Langley was a dream come true: they were going to improve the airplane, and at a one-of-a-kind research facility at that. "No one else in the country was doing this kind of work. It was so exciting it was unbelievable," says Axel T.

Mattson, who arrived at Langley in 1941 and retired from the Center as assistant director of External Affairs in 1974.

Key to Langley's research strength was an atmosphere that fostered exploration and initiative. Individuals were encouraged to find out what worked; if a device, modification or process was successful, it could then be incorporated onto an airplane for testing and verification. If, on the other hand, an idea had merit but its application was faulty or incomplete, then its originators went back to the drawing board to incorporate lessons learned and prepared for another try. For the newly minted college graduate, ready to make a permanent mark upon the world, Langley's greatest gift was the permission to fail. Learning by failure may appear a contradiction in terms, but such lessons are not easily forgotten. At Langley, the mistakes were just as important as the successes, for they sowed the seeds of future accomplishment.

Making test models in the 1930s.

"Hired fresh out of school with a minimum knowledge of aerodynamics and little practical experience of any kind, the majority of these early Langley researchers learned nearly everything on the job," writes *Engineer In Charge* author James Hansen. "Because they were so young, they had not learned that a lot of things could not be done, so they went ahead and did them."

Some unusual airplanes, like this experimental Custer channel wing, passed through Langley for evaluation. Designed to fly at very low airspeeds, this particular craft never made it into full production.

No matter how much latitude Langley's staff was given, when all was said and done, applied engineering was what the Laboratory was about. But researchers did not simply slap parts together to see what worked. The Langley way was one of systematic parameter variation: that is, meticulous, exacting variation of one component at a time to identify configurations that would produce the best results. Such a process took time, patience and cooperation above all else. At Langley, no researcher ever really worked alone. Successful application of aeronautical research demanded collaboration.

Theoreticians were essential members of the Langley staff. The task of these individuals was to chip away at the physics of flight with the hard, precise chisel of mathematics to explain and enlarge upon the results obtained in wind tunnels and in test flights. In the event that experimental results didn't agree with theory, either the experiment was repeated to verify the results or the theoreticians formulated new laws to explain the unexpected phenomena. But Langley theoreticians did more than scribble complex equations in notebooks. Their calculations led to the design of thinner and lower-drag wings, sturdier aircraft structures, better propellers and the first widely used deicing system, one that utilized the airplane engine's own exhaust heat.

For their part, Langley

engineers first used wood, then metal, to model new airplane designs. Laboratory researchers refined existing flight systems, improved engines and reworked original aerodynamic shapes. Because many of Langley's most talented engineers came to the Laboratory with little or no background in theoretical studies, it took them time to learn how to use theory to enlarge upon or improve their approach to engineering. Nevertheless, some of Langley's best work was done by such engineers, who managed to relate abstract theory to pragmatic aeronautical requirements in order to arrive at new engineering techniques or better devices.

Skilled technicians were also critical to Langley's ability to innovate. One of the most important factors considered by the Army and the NACA in site selection was the local availability of mechanics and technicians. Within an hour's car drive of Hampton there were numbers of workers skilled in wood, metal and concrete construction; in marine and automobile repair; in toolmaking and in the operation of electrical machinery. Such craftsmen were prized by Langley's professional staff because they provided the essential support services on which all NACA research programs depended. Without skilled technicians, research models could not

A one-twelfth scale model of an SBN-1 airplane undergoes tests in the 12-Foot Free Flight Tunnel in September 1940.

have been made, wind tunnels could not have been built or properly maintained and efficient day-to-day operation would have proven to be impossible.

The coalition of these groups, each with its own emphasis and strength, was the engine that drove Langley to research excellence. Those who went to the Laboratory for assistance were impressed by the staff's abilities and confident of receiving the best possible help. Said former McDonnell Douglas Aircraft Corporation official L. Eugene Root, when interviewed by historian Michael Keller in the early 1960s: "If you think the young guys at the NACA [couldn't make] your design … better, why, you have another think coming … No one company, or one individual, could have ever gotten it together [or] the facilities that were required to make the United States of America tops aeronautically. It would never have happened if it hadn't been for the wisdom of putting together these laboratories and giving young, driving, ambitious and damn smart … young men a place to be, a place to go and something important to do that was really fundamental to the country."

This model of a possible supersonic transport seems poised for takeoff from its wind-tunnel mount.

As intellectually nimble and technically shrewd were the Langley staff, they nonetheless needed first-rate laboratories and wind-tunnel facilities in which to do first-rate work. At Langley they got them.

Having the Right Tools

In 1901, to gather additional information on the performance of wing shapes, the Wright brothers built their first wind tunnel. It was a smallish contraption, a wooden box 6 feet long and powered by a two-bladed fan. The Wrights weren't the first to use wind tunnels in aerodynamic research—Briton Frank H.

A researcher is dwarfed by the fan blades of the Full-Scale Tunnel in this photo taken in 1931. In 1985, this facility—by then renamed the 30- by 60-Foot Tunnel—was declared a National Historic Landmark.

Wenham is generally considered to have originated the wind tunnel, in 1871—but their use of the device was central to the refinement and, subsequently, the success of their *Flyer.*

In simplest form, wind tunnels consist of an enclosed passageway—hence the term "tunnel"—through which air is pushed by one or more fans. Depending upon design, and whether outside air or another gas is used, the air flowing through wind tunnels has certain properties of speed, density and temperature. In order to mimic in-flight conditions and monitor a wide range of an aircraft's physical reactions to those conditions, researchers mount and instrument models (or, in certain instances, full-size craft) in the wind tunnel's heart, the test section. There, air or gas is made to flow around the (usually) stationary model. Many Langley tunnels take their names from the size of test sections: the 30- by 60-Foot Tunnel, for instance.

Throughout its history, Langley has taken pride in an extensive wind-tunnel complex, the largest of its kind—more than 40 wind tunnels are in operation at the Center—in the free world. Simply put, Langley wind tunnels are the linchpin of the Center's astronautics research program. There are specialized wind tunnels dedicated to a narrow range of investigations and wind tunnels in which a wide variety of experiments can be conducted. Langley's wind tunnels are small and large and are run at low, high and ambient temperatures. Some operate at many times normal atmospheric pressure, others at fractions of atmospheres. Models and shapes of airplanes, airplane wings, dirigibles, pontoons, submarines, satellites and spacecraft have all been evaluated in Center tunnels. Langley's wind tunnels are also durable, so much so that a handful remain from the Laboratory's earliest days, even if in substantially renovated form.

Today, Langley Research Center continues to upgrade and improve its wind-tunnel complex. Over the years, though, money has not always been easily available when the time came to renovate or replace tunnels. As it does today for NASA, the U.S. Congress held the purse strings for the NACA and carefully considered every request for new facilities. NACA officials appearing before Congressional committees were adept at explaining why funds were needed and exactly how the money would be spent. Still, being regularly grilled by committee was not something any NACA official relished. Nor did

In 1964 this HL-10 "lifting body" was mounted in the Full-Scale Tunnel for low-speed aerodynamic testing.

Congress routinely write the NACA a blank check for projects. Some were delayed, some denied outright. But there were ways to get around budget restrictions, as was demonstrated in 1937 when Langley decided to build a successor to one of its most productive wind tunnels, the Variable Density Tunnel, or VDT.

NACA officials felt that the expense of a VDT replacement could not be justified to congressional overseers; they simply would not understand the urgency. But the NACA Main Committee could obtain funding for a new tunnel if it was to be devoted to icing experiments. By 1937, many aircraft crashes traced to icing problems were attracting considerable public attention. Commercial airline operators were also clamoring for useful information on the subject. Thus it was that Langley began construction of an "ice" tunnel in May 1937.

Former NACA Langley employee W. Kemble Johnson recalled in a 1967 interview his role in the project:

> We built it from scratch—I mean, we were poor people. At Fort Eustis [a nearby Army base] we scrounged steel, trusses and columns that had been torn down and were laying in the weeds with trees practically growing through them. Because they were twisted and out of shape I had burners and welders come in.... [They] straightened out the trusses... took columns... cut the ends off and welded another column to them to get the height... For less than $100,000 we built the whole building and wind tunnel and the works....

These vacuum spheres, now part of Langley's Hypersonic Facilities Complex, have been dusted by a light fall of snow in this 1969 photo.

The ice tunnel's insulation came courtesy of the U.S. Navy. Surplus Navy life preservers were obtained, and high school students cut the vests open to fluff out the insulation before it was applied. The refrigeration system consisted of dry ice, automobile carloads of which were unloaded by the same intrepid students. The first operational run of the tunnel came at night and presented a rather eerie sight. An opaque dry-ice fog hung above the floor and, Johnson reported, "The light would shine down on us and we'd walk around with just our heads sticking up. On top of [that fog] layer ... was about a half-inch thick layer of mosquitoes with their jaws open." It was, he concluded, "a very weird thing."

The ice tunnel was only used for a brief series of experiments before conversion to a low-turbulence wind tunnel. Eventually, parts of the ice tunnel were used in the development of the test section and entrance cone of the Low-Turbulence Pressure Tunnel. In the battle of wits and pocketbooks, at least in this instance, the Laboratory had emerged the victor.

If there were ways to get around funding bottlenecks, there were also ways to get around research restrictions in the wind tunnels themselves. Langley wind-tunnel studies haven't always been officially approved; the practice of "bootlegging"—unauthorized, if imaginative, research—has occurred over the years. Two of the more ambitious Langley bootleggers were Arthur Kantrowitz and Eastman Jacobs. In 1938 the pair undertook what is believed to have been the world's first attempt to construct a nuclear fusion reactor. The project was abruptly cancelled, however, when discovered one day by Dr. George Lewis, the NACA director of research, visiting from NACA headquarters in Washington.

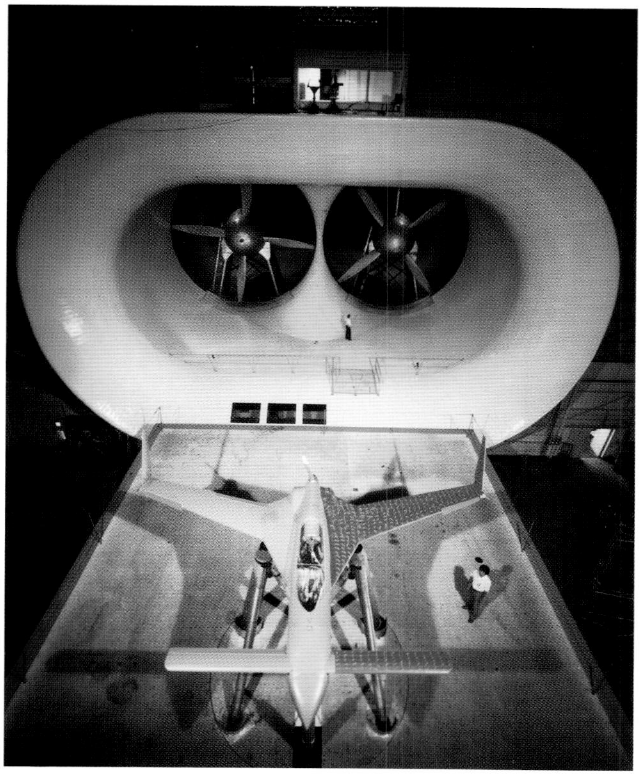

Tunnel work has also presented a degree of physical risk. High pressures can lead to explosions, or structural failure of fan blades can tear a building apart. In one instance that occurred in the late 1950s, two technicians were blown right out of a tunnel into a nearby swamp when pressurized air was improperly vented. Fortunately, both survived. In another mishap around the same time, a test run of a high-temperature tunnel that used superheated pebbles resulted in a score of minor fires when the pebbles were inadvertently ejected outside of the tunnel. Paint was even burned off nearby cars.

Barton Geer, who retired in 1981 as Langley's director for Systems Engineering and Operations, was introduced to the perils of wind-tunnel research in 1942. As a recently arrived junior engineer, he was sent to work in the 19-Foot Pressure Tunnel. One day, Geer was instructed to take pressure and humidity readings in the tunnel's test section. In order to do so he had to enter an airlock. But, says Geer, "In the early years, we didn't think about safety like we do now. So the fellow who put me in there went home and forgot all about me. I didn't know how to work the airlock to get out. I was thinking, 'What's going on here? What's my wife going to think?' Fortunately, around midnight he said, 'My gosh—Bart's still in there!' So he came back and got me out."

Advanced-concepts aircraft, like this Vari-Eze designed by aeronautical innovator Burt Rutan, are regularly evaluated in Langley wind tunnels.

In recognition of the Center's wind-tunnel contribution to aeronautical science, three Langley tunnels were declared National Historic Landmarks in 1985 by the U.S. Department of the Interior. They are the Variable Density Tunnel, 30- by 60-Foot Tunnel and 8-Foot High-Speed Tunnel. Two other facilities—the Lunar Landing Research Facility and Rendezvous Docking Simulator—were also proclaimed Historic Landmarks.

Take human ingenuity out of the picture and Langley's wind tunnels are nothing more than an expensive amalgamation of steel, bricks, mortar and sophisticated equipment. But allow for human drive and creativity, as Langley has done, and these state-of-the-art "tools" can be seen for what they are: among the wisest capital investments the federal government has ever made.

A technician prepares to unlatch a door built into the guide vanes of the 16-Foot Transonic Tunnel. The vanes—which here form an ellipse 58 feet high and 82 feet wide—make possible a smooth and uniform flow of air through tunnel passageways.

Beyond Brainpower

No one factor can be isolated as the sole agent responsible for Langley's technological prominence in aeronautics. There does seem to have been something of a Langley cultural "formula"—a mix of sharp intellect, curiosity, humor, enthusiasm, competitiveness, personalities and personality clashes—that enabled aeronautical researchers to do their best work. "What impressed me most about Langley," says Donald Hearth, director from 1975 to 1984, "and what made Langley so different, were the people. They were extremely creative, highly loyal, very competent, always worked well together, particularly when the challenge was great, and believed that they could do almost anything." Exceptionally able hands also appear to have held the management reins. Many veterans credit men like George W. Lewis, the NACA's first director of research, and Langley's engineers in charge with setting the Center on the proper course and guiding it through the shoals of project selection and program expansion. (By 1960, with the appointment of Floyd Thompson, the title of the individual overseeing Langley was changed to Center Director.)

Regardless of how it exercised its expertise, Langley had enough to spare. Langley exported its organizational and engineering talent, first to Langley's daughter NACA laboratories and, later, to NASA's Washington headquarters and to emerging space centers. In the opinion of some, it is not overstating matters to describe Langley organizational know-how as crucial to the success of the U.S. manned space program. "One of our primary 'products' has been people: leaders, really, in the aerospace field," maintains Langley Research Center Director Paul Holloway. "We sent groups to found other centers, like Dryden, Lewis, Ames and Wallops. Many went on to Washington and played major roles in agency management. In '61, '62, a group left here to start Johnson Space Center—totally from scratch."

Langley's engineers might have been bright and creative, and its leaders adept at technology management, but the Laboratory was not immune to the petty suspicions that inevitably arise when a small town becomes the home of those thought to be outsiders. In the early years of Langley's existence there was something of a culture clash between the local populace and the professional Laboratory staff. A significant percentage of that staff came from more populous areas in the North and Midwest, where amusements were many and easy to come by. Hampton was Southern, rural, isolated, a place to make fun of but not a place in which to have fun. The clannish Hamptonians were made uneasy by the brash confidence displayed by the NACA "Yankees." Matters weren't improved

Members of the Laboratory's model-airplane club, the "Brain Busters," founded in 1942.

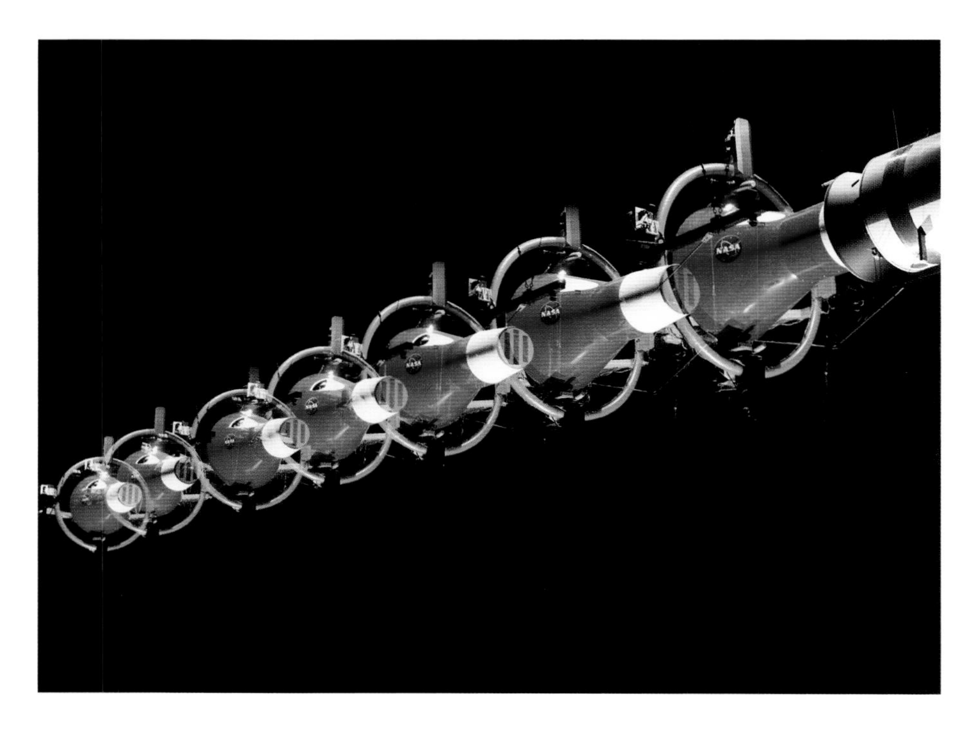

A multiple-exposure photograph of Langley's Rendezvous Docking Simulator, used by astronauts to train for the Gemini program. (Designated a National Historic Landmark in 1985.)

On June 26, 1959, then-Langley-researcher Francis Rogallo examined the "Rogallo Wing" in the 7× 10-Foot Tunnel. Originally conceived as a means of bringing manned spacecraft to controlled, "soft" landings, Rogallo's concept was avidly embraced by later generations of hang-gliding enthusiasts.

when, in response to their cool reception, some Langley researchers didn't hesitate to tell the locals what side of the streetcar they should get off.

"Hampton was a sleepy fishing town. As the saying goes, you could fire a cannon down Main Street at 9:00 p.m. and not hit anyone," remembers Don Baals, who came to work at Langley in 1939 and who retired in 1975 as assistant chief of the Full-Scale Research Division. "The Hampton people viewed these [NACA] people with a degree of trepidation. But the problem was solved when the young men married into the local families."

For years the phrase "Nacka nut" (Nacka was the spoken form of NACA) was heard around Hampton and surrounding environs. The detail-oriented Langley engineer, it was said, would venture into hardware stores and ask that lumber be cut to the nearest sixteenth of an inch, a ridiculously precise amount. Or a hapless appliance salesman would be waylaid and asked to detail the manufacturer's specifications for a vacuum cleaner, including the number of revolutions per minute made by its electric motor.

Once, or so the story goes, a Laboratory engineer bought a hand-cranked ice-cream maker from a local hardware store. The appliance came with a rustproof guarantee. Three weeks after the purchase, the engineer returned; the maker was a rusted ruin. The store owner replaced the original with another, also guaranteed against rust. Another three weeks went by and the engineer returned, with the second maker in the same condition as the first. Again a free replacement was provided. Two weeks later, the engineer was back, this time with a third rust-encrusted ice-cream maker. Incredulous, the hardware store owner asked the engineer exactly how he made his ice cream. The engineer replied that he would make no ice cream until he was satisfied that the maker was really rustproof. Therefore, the engineer added, he had filled the makers up with salt water and let them sit in his back yard. Thus far, none had passed the test.

The owner promptly refunded the engineer's money and told him never again to think about buying an ice-cream maker—or anything else, for that matter—from that particular hardware store.

On balance, and as time passed, negative encounters between Langley employees and Hampton residents became far less frequent. The locals grew accustomed to the accents and habits of the young researchers that came to Hampton from all over the country. Many rented rooms in area boardinghouses, ingratiating themselves slowly but surely into the community's daily routines. Over time, familiarity bred contentment.

Apart from their standing in Hampton at large, those working at Langley Laboratory made a point of enjoying themselves among themselves. Laboratory staff developed a lively social circuit: a club for model-airplane enthusiasts; touch football, basketball and baseball teams; rounds of parties; regular outings to nearby beaches; frequent dances and periodic gatherings of every sort. Some were talented musicians and delighted their colleagues with prowess on the piano or other musical instruments. Others were lusty singers and one or two were able amateur magicians.

While the Langley staff were serious about work, they were serious about fun, too. John Becker began his work at Langley Laboratory in 1936 and retired in 1975 as chief of the High-Speed Aerodynamics Division. In his book *The High-Speed Frontier*, Becker recalls that, even during World War II, sometimes a good diversion was nothing more than a well-thought-out practical joke:

Revelers prepare for a human wheelbarrow race at the Langley Laboratory annual picnic, held at Buckroe Beach in the late 1920s.

> The staff relaxed through all of the usual sports and social events with little apparent effect of wartime pressures. Five of us had formed an informal golfing group.... [My boss John] Stack had never played before and had no clubs of his own, but we offered to lend him an old bag with a broken strap and some of our spare clubs.... [Henry] Fedziuk, who was the chief humorist of the group, had often been the butt of Stack's practical jokes and saw here a welcome chance to turn the tables.
>
> With enthusiastic help from some of the rest of us he lined the bottom of Stack's bag with some 10 pounds of sheet lead. We also made sure the bag had a full complement of clubs, and we told Stack that caddies were used only by the rich and decrepit. By the start of the back nine, with a score card showing well over a hundred in spite of considerable cheating, Stack was seen to start dragging the bag along behind him....
>
> His expletives [became] louder and more colorful, and a short time later he discovered what had been done. Understandably, he always examined his equipment very suspiciously at subsequent sessions.

During the same event, onlookers cheer their favorites during the barrel-joust competition.

The spirit of camaraderie extended to the labs, where cooperation and collaboration was seen

Miniature Apollo concept model casts a larger-than-life shadow in this 1964 photo, taken in the slotted-throat 16-Foot Transonic Tunnel.

as a virtue. But there was also a good-humored rivalry. "There was enormous technical competition between the divisions at Langley," remembers Israel Taback, who arrived at Langley in the early 1940s and who, upon retirement from Langley in 1976, was the chief engineer on the first project to ever soft-land a robot probe on Mars. "People would fight with each other over technical details. That was all very healthy. The end result was a battle of ideas—ideas that had merit tended to float to the surface. The good ideas won."

That Langley's was an environment suited to achievement was underscored by the multitude of national and international awards won by staff members over the years. Perhaps none was more prized than the Collier Trophy, named in honor of publisher, sportsman and aviator Robert J. Collier. Since 1910 the Collier has been awarded annually for the greatest achievement in American aeronautics (and, recently, for astronautics achievement as well). Langley researchers have been thus acknowledged five times: in 1929, for a low-drag engine cowling; in 1946, for research on airplane icing; in 1947, for supersonic flight research; in 1951, for development of the slotted-throat transonic wind tunnel; and in 1955, for the transonic area rule.

The point can be made that, since so relatively little was known about the specifics of flight, it was almost guaranteed that Langley researchers would unearth something that could be productively applied to the flying of airplanes. But nothing is ever guaranteed. That Langley Laboratory achieved what it did is tribute to the talent and drive of its staff and the savvy of NACA officials and supervisors who knew when and how to exercise control. Langley's ultimate contribution was not that of the manufacturer, for the Laboratory would never build airplanes. Rather, Langley donated its intellectual currency to the advancement of aircraft; its true value to the aeronautical industry and the nation was that of aeronautical trailblazer.

In time, later generations of flying machines would surpass the Wright *Flyer* in the same way a modern automobile outstrips a primitive two-wheel cart. Prowess in the atmosphere led directly to success in space. Ever more sophisticated craft would be developed, craft that could "slip the surly bonds of Earth." Yet close to 7 decades would pass before humankind was able to make the long leap from a wind-swept Carolina beach to the Moon's Sea of Tranquillity. During that time, Langley Research Center would contribute to ventures that would have appeared preposterous to even the most visionary of 19th century aeronautical pioneers.

Close to 7 decades would pass before humankind was able to make the long leap from a wind-swept Carolina beach to the Moon's Sea of Tranquillity.

KING

Perfecting the Plane
1917–1939

High above the mud, blood and gas attacks of World War I trench warfare flew remarkably flimsy craft that were, by the standards of the day, a stunning technological advance. Here was proof positive that the airplane was an invention with which to reckon. The plane was no longer a comic extravagance or adult toy, for the outbreak of military conflict mandated for it a darker purpose—that of a powerful agent of war. As the aircraft of the warring powers sparred with one another in the world's first dogfights, it was quite clear that the airplane's role had been forever altered.

Clad in a fur-lined leather flying suit with oxygen facepiece, NACA test pilot Paul King prepares to take to the air in a Vought VE-7.

At war's end, the European rail system in shambles, the role of the airplane was again expanded, this time as an instrument of peace and commerce. The private-sector aviation industry slowly began to grow, led by individuals determined to find a profitable niche in the transportation of people and goods. There was certainly plenty of equipment and skilled workers, for war had provided an abundance of aircraft and pilots willing to fly them.

Within three months after the November 1918 armistice, commercial aviation began in Germany as Deutsche Luftreederei inaugurated passenger-carrying service; that same year daily flights between London and Paris commenced. The first passenger flights between U.S. cities followed in 1920 and, by 1925, regular air-freight service between Chicago and Detroit had been established.

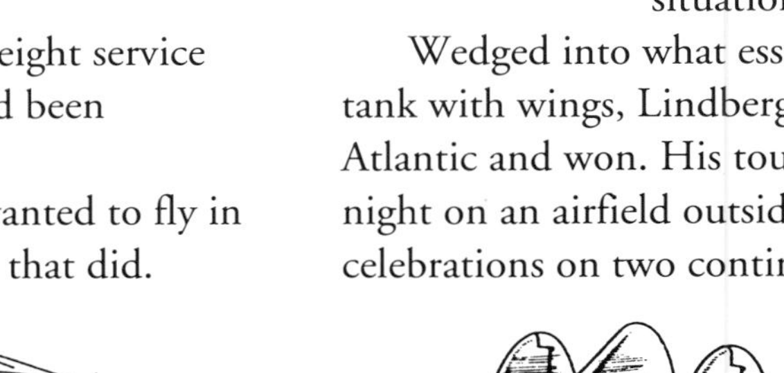

Stunt flyer Lincoln Beachy and his Curtiss "Looper," in which he performed his trademark loop-the-loops throughout the country.

Everyone, it seemed, either wanted to fly in an "aeroplane" or knew someone that did. Enthusiasts predicted that the airplane's exciting childhood would usher in a brighter, faster future. Soon, speculated these starry-eyed proponents, there would be a personal airplane in every garage. It was simply a matter of time. But however accustomed the general public was becoming to the drone of aircraft engines overhead, to the sight of goggle- and leather-clad aviators, and the notion of sending or receiving "air-mail," in physical and economic terms actual flight remained a relatively risky business. Crashes were not uncommon. With the exception of a handful of hardy commercial carriers that pampered well-to-do clients and ferried mail under contract, few American companies found profit in aviation. The federal government and the military remained the primary buyers of new aircraft and the sponsors of most aeronautical research. Fortunately for the commercial aviation industry, the nonstop transatlantic flight in 1927 of aviation pioneer Charles Lindbergh—coming as it did almost a quarter century after the Wright *Flyer* rose above the sands of Kitty Hawk—dramatically changed the situation.

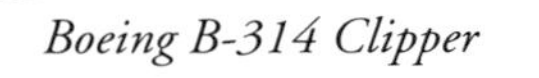

Boeing B-314 Clipper

Wedged into what essentially was a flying gas tank with wings, Lindbergh dared the wide Atlantic and won. His touchdown on a cool May night on an airfield outside Paris set off wild celebrations on two continents. But Lindbergh's gutsy accomplishment was more than a personal triumph, for it proved once and for all that the airplane could conquer great distance. "Lucky Lindy's" success drew worldwide attention to the airplane's ocean-crossing potential and, not incidentally, inspired an entire generation of young, would-be aeronautical engineers and aviators. By the late 1930s, coast-to-coast air service was a routine fact of life and "flying boats" were beginning regular treks across transpacific routes.

Goggles at the ready, this Langley test pilot and engineer conducted research business high above the ground.

Just after World War I, the bulk of Langley's research still aimed squarely at solutions to problems of specific concern to the military. But by the late 1920s, as the role and importance of commercial aviation increased, so had the time the Laboratory devoted to study of aeronautical items of interest to the private sector. Fortunately, what had been learned in Langley's studies of military aircraft design could usually be applied, with minor modification, to civil aviation. (By the late 1930s military and private-sector interests were diverging, as the military became interested in higher speeds and altitudes while commercial carriers emphasized safety and efficient operation.)

By 1927, aeronautical research at the NACA Langley Laboratory was in full swing. Extensive theoretical and experimental work was being done on lighter-than-air (LTA) craft—known popularly as airships or dirigibles—in tandem with the U.S. Army. Langley personnel conducted tests to determine takeoff, landing and docking procedures, and assisted in speed and deceleration measurements. As a result, writes *Engineer In Charge* author James Hansen, many Langley flight researchers became outspoken advocates of airships:

The device mounted at the tower's apex made a turning-radius measurement of the Navy dirigible U.S.S. Los Angeles *in this photo taken in 1928.*

> It was not clear at all to them or to anyone else at the time that the airplane would win out over the airship, let alone as totally as it soon did. Airplanes of the early 1920s were slow and small—an aerodynamicist who favored airships over airplanes even went to the bother of "proving" that airplanes larger than those of the day could never be built. LTA advocates believed correctly that airships had enormous unproven capabilities: they were not much slower and could carry many more passengers in far greater comfort than airplanes, most of which still had open cockpits; they were much more forgiving than airplanes during instrument flight; and with their extreme range and low operating cost, they could be used not just as military weapons but also for transportation of heavy commercial and industrial loads.

In 1924, this modified Ford Model A and a "Huck starter" were used to crank aircraft engines to life.

In aviation's youth, instrument panels and controls were models of simplicity—at least compared with those of modern aircraft. The cockpit illustrated is a Fairchild FC-2W2 from 1928.

1. Fuel Tank Selector Valve
2. Compass
3. Airspeed Indicator
4. Starter Switch
5. Fuel Pressure Gauge
6. Altimeter
7. Master Switch
8. DC Ammeter
9. DC Voltmeter
10. Manufacturers Plate
11. Engine RPM Gauge
12. Engine Oil Pressure Gauge
13. Engine Oil Temperature Gauge
14. Instrumentation Switches (wing attitude, flight path angle, and dynamic pressure)
15. Magneto Switch
16. Spartan Throttle Quadrant
17. Throttle
18. Mixture Control
19. Propeller RPM Control

Unfortunately, the infamous accident on May 6, 1937, that destroyed the dirigible *Hindenburg* as it attempted to dock in Lakehurst, New Jersey—23 crew and 13 passengers lost their lives when the airship burst into flames—also resulted in the economic collapse of the 20-year-old LTA passenger-carrying industry.

Simultaneous with its LTA studies, Langley continued aircraft research. New models manufactured by such companies as Curtiss, Martin, Sperry, Vought, Douglas and Boeing underwent evaluation at the Laboratory. Langley's work revealed, and contributed to, an improving airplane: one that was becoming safer, faster, stronger, easier to handle. But the plane was far from perfected. Designing the best possible aircraft proved more often than not a trade-off between desirable characteristics, such as speed and range. Moreover, the forces that permit and constrain flight are complex. Understanding them required time, determination and ingenuity.

On the Job

Langley's first building erected was, by modern standards, a modest affair. Built by the New York City firm of J. G. White Engineering

Corporation at a cost of $80,900 in 1917-era dollars, the structure contained administrative and drafting offices, machine and woodworking shops and photographic and instrumentation labs. The Laboratory's first wind tunnel was separately housed in a small brick-and-concrete building. By 1922 the Langley complex had grown to include two wind-tunnel facilities, two engine dynamometer laboratories and a large airplane hangar. Research was being conducted on better flight instrumentation and ways to reduce aerodynamic drag, increase lift, boost propulsion efficiency and improve structural integrity.

For more than a dozen years after its official formation, the Langley professional staff numbered less than 100, a figure that was not to be surpassed until 1930. (By 1927 support, or "nonprofessional," staff had grown to 104 individuals.) That this relatively small complement would again and again produce top-notch results might have been due to the balance between structure and independence, a process that author James Hansen terms "careful bureaucratic restraint [and] research freedom." At Langley there was great institutional reluctance to announce results of studies until researchers and their superiors were confident that those results would bear up even under the toughest scrutiny. Researchers were therefore free to work creatively on novel ideas without the fear of preliminary reports building up too much industry anticipation of and pressure for future advances.

A Curtiss Jenny trails a pitot-static tube, a device used to calibrate airspeed.

The honeycombed, screened center of this open-circuit air intake for Langley's first wind tunnel insured a steady, nonturbulent flow of air.

The Langley working atmosphere was one of informality. Everyone knew everyone else, and the most junior could personally get acquainted with the engineer in charge. There was an organizational chart, but it was seen more as a necessary evil. "Titles were tall cotton. People were not here for self-glorification," says William D. Mace, who came to Langley in 1948 and who retired in 1989 as director for Electronics. "The thing that held folks together out here was their common interest: the ability to do first-class aeronautics research. The fact is, Langley produced. If it hadn't, it would have disappeared."

In the first decades of its existence Langley management did its best to keep a safe distance between the Laboratory and bureaucrats in the nation's capital. John Becker, writing in *The High-Speed Frontier*, observes that the Langley of the 1930s did not think of itself as part of the federal bureaucracy. Langley was "spiritually and physically separated from Washington. The youthful staff had been largely handpicked in one way or another to form an elite group unique in the federal system... [There was] a beneficial sense of family."

As in any family, at Langley there were occasional disputes, personality clashes and struggles over the nature and extent of research programs. Whatever problems arose were refereed by management, a group small in number but fiercely dedicated to Langley's flight-research mission. Managers didn't mind dirtying their hands; indeed, many relished it. That Laboratory management was of the hands-on variety soon became evident even to the most junior engineer. John Becker writes of his introduction to the Langley management style while preparing an experiment in the 8-Foot Tunnel:

> One night during my second week on the job, just before I closed the airlock doors at the entrance to the test chamber for a test run, an unusual-looking stranger dressed in hunting clothes came in and stood there watching my preparations. [My supervisor] had advised me not to allow visitors in the test chamber during a high-speed run primarily because the pressure dropped quickly to about two-thirds of an atmosphere, the equivalent of a 12,000 foot altitude.
>
> Assuming that the visitor had come in from one of the numerous duck blinds along Back River, I said firmly, "I will have to ask you to leave now." Making no move he said, "I am Reid," in such ponderous and authoritative tones that I quickly realized it was Langley's Engineer In Charge whom I had not yet met.
>
> No one had told me that Reid, who lived only a couple of miles from Langley Field, often came out in the evening, especially when tests of electrical equipment were being made (he was an electrical engineer)....

The white marble tabletops of the Langley lunchroom were a boon to researchers, who often sketched "tabletop" equations. Their marks could be easily erased with a hand or napkin.

Today there is much talk about how to improve the efficiency of public and private enterprise. The intent is to "flatten the pyramid," to eliminate unnecessary layers of management in awkward command-and-control systems, systems that centralize power, reward bureaucracy and stifle creativity. From the very beginning Langley had few such problems. Laboratory management encouraged the free flow of ideas, whether they came from a grizzled veteran or a just-hired. If an idea had merit, a junior engineer could approach his superiors without fear of reproach. If the idea was successfully adopted, the individual proposing it would receive full and proper credit.

There was a brisk exchange of ideas at Langley, in discussions not just limited to the lab. Some of Langley's best work was done while researchers were out to lunch—literally. Most of the professional staff assembled on a daily basis in the second-floor lunchroom of the Laboratory's administration building. There, "plate" lunches could be had for 25 or 30 cents (35 cents on days steak was served). The lunch tables had white marble tops, a feature which was a great boon to technical discussions. Researchers could and did draw curves, sketches and equations directly on the table during animated exchanges, marks that could easily be

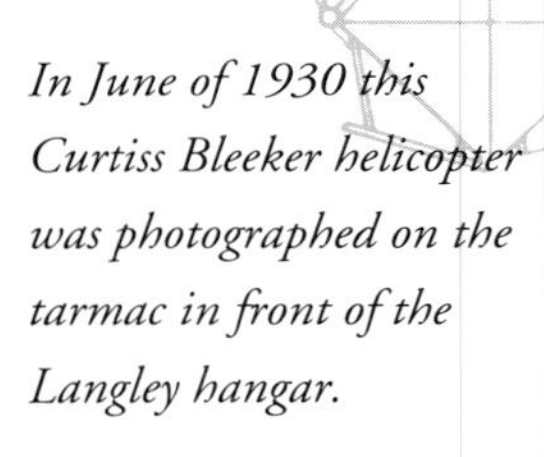

In June of 1930 this Curtiss Bleeker helicopter was photographed on the tarmac in front of the Langley hangar.

erased with a hand or napkin. "It was exciting and inspiring for a young new arrival to sit down in the crowded lunchroom and find himself surrounded by the well-known engineers who had authored the NACA papers he had been studying as a student," John Becker writes in *The High-Speed Frontier*. "There were no formal personnel development or training programs in those days, but I realize now that these daily lunchroom contacts provided not only an intimate view of a fascinating variety of live career models, but also an unsurpassed source of stimulation, advice, ideas and amusement."

Two mechanics measure and record wing ordinates on a Curtiss Jenny airplane.

But however challenging and intellectually exciting Langley's aeronautical research was, it was far from glamorous. Young engineers worked long, hard hours. The recently hired paid their dues by laboriously plotting by hand the data collected from wind tunnels, supervising the mounting of models, turning valves, watching gauges, and generally making sure that everything was shipshape before wind-tunnel tests were run. The work was routine, even boring, but for engineers in love with aeronautics, the rigors of the work paled in comparison to what could be, and was, learned.

There was a certain price to pay for the Langley can-do reputation. As the Laboratory attracted more national attention, it began to lose some of its best and brightest to the booming private sector, which beckoned with higher salaries and hard-to-refuse research opportunities. Between 1920 and 1937, 37 professional staff left Langley for aeronautical careers elsewhere. Considering Langley's size, such a loss was significant. As James Hansen notes, though the personnel losses may have delayed the successful execution of a few NACA research projects, the larger American aeronautics effort probably benefitted from them. Langley provided a training ground for some dozens of aeronautical experts, and an apprenticeship there was excellent preparation for a university career or a job with a major aircraft manufacturer.

Many who came to work at Langley intended to stay but a few years and then move on. However, not all who thought of the Laboratory as a professional steppingstone followed through on their original intentions. Langley's character, its sense of community, its technical culture, its strong sense of self and mission, the sheer number of aerodynamics challenges that confronted its staff and the chance to make a difference: these were persuasive arguments that convinced not a few to stay put at the Laboratory. Certainly, for those who elected to remain, there would be no shortage of interesting projects.

ARMY
NINTH ANNUAL AI
EXECUTIVES AND E
NATIONAL
LANGLEY FIELD,
VIEW SHOWS

Gathered together in the only facility big enough to hold them, attendees at Langley's 1934 Aircraft Engineering Conference pose in the Full-Scale Tunnel underneath a Boeing P-26A Peashooter. Present, among other notables, were Orville Wright, Charles Lindbergh and Howard Hughes.

Breaking Through

Exactly how it is that human beings make an intuitive leap from half-baked idea to sound concept remains something of a mystery. What isn't mysterious is that one's chances for making the right conceptual connections increase the longer one works at it. Perhaps Thomas Edison said it best when he described genius as consisting of 1 percent inspiration and 99 percent perspiration. Hard work was the norm at Langley, but it was work which researchers eagerly embraced. Motivating them was a feeling similar to that felt by pioneers crossing unexplored territory: anticipation, enthusiasm, a sense of pending accomplishment. "Langley engineers knew they were making fundamental contributions toward understanding how an airplane flew," says John C. Houbolt, who came to Langley in 1942 and who retired in 1985 as the Center's chief aeronautical scientist (13 of those years were spent in the private sector). "Langley was breaking through, on the frontiers of technology."

The Variable Density Tunnel arrives by rail in 1922 from the Newport News Shipbuilding & Dry Dock Company.

What Langley's young engineers did in the 1920s was whittle steadily away at a block of assorted aeronautical problems. One of the most intractable dilemmas was that of speed: how to make planes fly faster, while maintaining acceptable safety standards and operating efficiencies. Langley's high-speed research, begun in the '20s, continued even as speeds geometrically increased compared with those of the early years. Laboratory researchers also worked on small-scale projects with precise objectives, like the instrument program that

June, 1922: Workmen in the patternmakers' shop manufacture and assemble a wing skeleton that will be used for in-flight pressure distribution studies.

For all the progress being made in airplane flightworthiness, designers still had an incomplete understanding of the interaction between the aerodynamic forces acting on an aircraft and the aircraft's structural response to those forces. Two areas of particular concern to researchers were aeroelasticity—the tendency of aircraft to twist and bend while in flight—and flutter—destructive harmonic vibrations of an aircraft structure reacting to an airflow. Flutter is thought to have been partially responsible for the 1931 in-flight breakup of a Fokker trimotor, causing the deaths of famed Notre Dame football coach Knute Rockne and six others. Theoretical models developed at Langley provided a means to calculate the forces causing flutter, thereby allowing Laboratory engineers to suggest ways to structurally strengthen the most flutter-susceptible parts of an aircraft.

In 1940 Langley formally dedicated the Structures Laboratory, its first facility devoted strictly to the study of aircraft structures. There, researchers worked on ways of making an airplane's metal skin stiffer and stronger and examined methods to internally brace the weakest areas. Fatigue—the tendency of metal structures to buckle or break after repeated use—was also investigated. Fatigue experiments done at Langley and elsewhere eventually led to "rip-stop" designs that minimized crack propagation (the tendency of a small tear to become a catastrophic rip) by reinforcing an airplane's frame at key points.

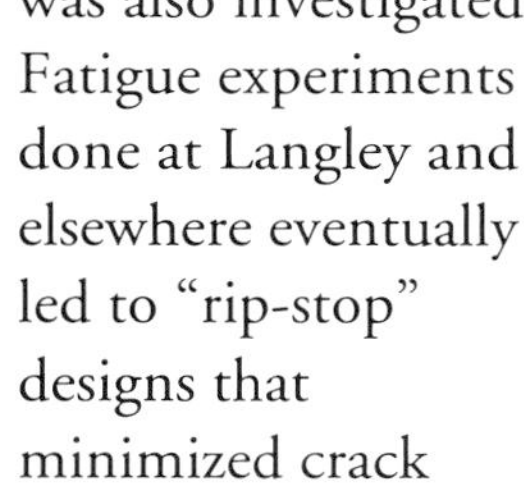

Douglas DC-3

If there was any one airplane that epitomized the design revolution of the 1930s, it was the Douglas DC-3 transport. Langley had an active role in developing or evaluating the DC-3's aeronautical innovations, which included internally braced wings, wing flaps, retractable landing gear, cowled engines, controllable-pitch propellers, a geared supercharger and all-metal,

At work in the metal shop making engine cowlings.

Langley's seaplane towing facility (right) and the Full-Scale Tunnel (left) were photographed in November of 1930.

stressed-skin construction.

The DC-3, which first flew in December 1935 and was in airline operation by the summer of 1936, was sufficiently large to carry 21 passengers. With this number of passengers and a cruising speed of 185 mph at 10,000 feet, the airlines for the first time had an aircraft with operating costs sufficiently low so that money could be made without complete dependence on revenue from airmail contracts. The craft was, as some pilots described it, "one tough bird": although easy to handle, the DC-3 could absorb structural punishment and keep on flying. By 1940, the existing fleet of DC-3s had flown 100 million miles, carried nearly 3 million passengers and had become the dominant airplane of its time.

Langley's contributions to the development of an aircraft such as the DC-3 would not have been possible without additional state-of-the-art research facilities which, by the early 1930s, were becoming operational at the Laboratory. In 1931, for instance, the Full-Scale Tunnel joined Langley's wind-tunnel roster. Into its 30- by 60-foot test section a modest two-story house could comfortably fit; most aircraft of the era could easily be accommodated as well. (So useful has the Full-Scale Tunnel been to Langley that it exists still, refurbished and renamed, as the 30- by 60-Foot Tunnel). By mid-1931 a hydrodynamics facility—known at Langley as the Towing Tank—was also put into operation. Originally 2,000 feet long, it was later extended to 2,900 feet and was used primarily to determine the performance characteristics of various hull designs for seaplanes and amphibious vehicles. By towing model hulls through the water from a standing start to a simulated takeoff speed, researchers could suggest changes in or improvements to basic designs.

By 1935, the 15-Foot Spin Tunnel had been built and by the late 1930s a series of high-speed tunnels—the 11-Inch, 24-Inch and 8-Foot—

were completed. The 24-Inch High-Speed Tunnel was especially productive: by 1939, tests of airfoils therein had led to the design of the propellers that powered the 400+ mph American fighters that ruled the European and Asian skies in the last years of World War II.

In 1936, the 8-Foot High-Speed Tunnel began operations. There new aircraft models could, for the first time, be evaluated at speeds in excess of 500 mph. Based on pioneering investigations conducted in this facility, researchers were able to delineate the specific stability-and-control problems encountered in high-speed dives. Practical aircraft products to result from the studies included a dive recovery flap, high-speed low-drag engine cowlings, a new family of air inlets for jet-propelled aircraft and designs for 500+ mph propellers.

Early in 1937, a contract was awarded to begin construction of the 19-Foot Pressure Tunnel, which became operational two years

An interior view of the seaplane towing channel, where a variety of hull and pontoon shapes were evaluated.

A Douglas YO-31A aircraft is set up for tests in the Full-Scale Tunnel in late May of 1932.

The Fred Weick-designed W-1 with tricycle landing gear, in the Full-Scale Tunnel, in March 1934.

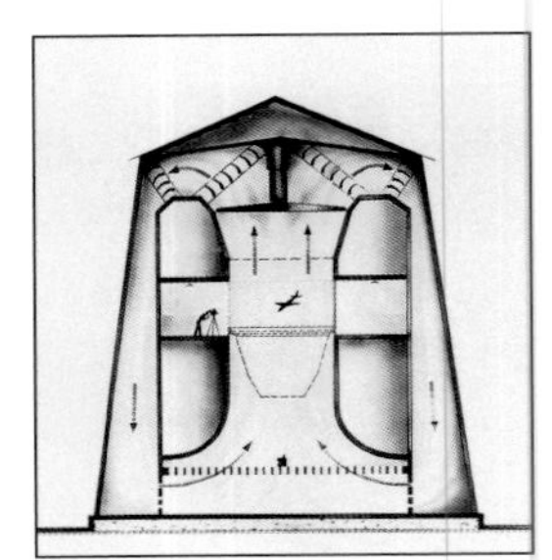

A cross-sectional interior view of the 20-Foot Spin Tunnel. Models were launched Frisbee-fashion into the ascending airstream.

An exterior view of the 20-Foot Spin Tunnel.

later. There, under 2.5 atmospheres of pressure, were examined various aircraft control and flap systems as well as designs for a number of World War II-era airplanes. When more advanced tunnels were developed later, the 19-Foot was assigned to research in aircraft aeroelasticity and high-speed flutter. Eventually the facility found new life, with Freon gas as a test medium, a new 16-foot test section and a new name: the Transonic Dynamics Tunnel.

By the late 1930s, Fred Weick, of NACA cowling fame, had devised an effective, if unconventional, "tricycle" landing gear, improving upon a design introduced by the Wright brothers. Weick positioned a single strut with attached tire under a plane's nose and a wheel under each wing. Because the two main wheels were behind the plane's center of gravity and the nose wheel was steerable, it was far easier to taxi and land an airplane. Pilots approved of the improvement in visibility—the plane sat more level on the runway—and passengers were grateful that they no longer had to scramble up and down inclined aisles. Beginning with prototype versions in the late 1930s, by the late 1940s nearly all U.S. commercial and military aircraft employed the tricycle concept, or a version thereof.

Martin M-130 (China Clipper)

By Langley's 22nd birthday, in 1939, the world had been made a different place by the advent of ocean-crossing airplanes. The tyranny of distance had been overcome and travelers were crossing the Atlantic and Pacific Oceans in increasing numbers. By contemporary standards, air travel was slow and time consuming—a trip from London to New York on Pan American Airways' "flying hotel," the B-314, took 23 hours—but stylish and comfortable nonetheless. The introduction of Pan Am's "China Clippers" and the construction of island-based resorts and refueling depots made passenger-carrying transpacific flight feasible, even enjoyable. The airplane had become an intercontinental, paying proposition.

"Many people knowing aviation considered that [commercial] transoceanic flight would forever be impossible," remarked famed aeronautical-design pioneer Igor Sikorsky in an interview conducted in October 1971. "[But] the NACA by [its] work ... certainly helped to produce these ships and certainly helped to bring and keep America in the first place in commercial aviation. Military too, but commercial aviation was definitely first because of the very excellent scientific work which this organization produced."

The airplane had ascended to youthful prominence directly after the First World War as a carrier of people and goods. Its vigorous adolescence in the 1930s was marked by

Built in 1935, the 8-Foot High-Speed Tunnel (named a National Historic Landmark in 1985) was the world's first 550-mph wind tunnel large enough to investigate shock-wave problems on complete aircraft models, cowlings and propellers. The concrete walls of the igloo-like test section (center) were 1 foot thick.

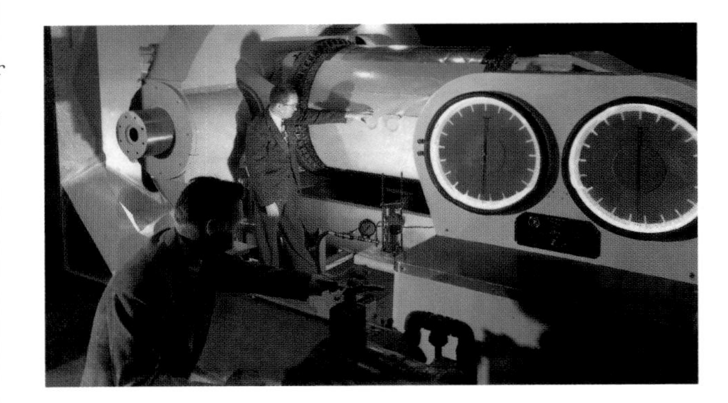

Structures research at Langley included studies of methods to prevent failures of pressurized fuselages.

substantial design changes and the further maturation of globe-girding commercial markets. But political conflict would again drive technological change. By 1940 the planet was embroiled in yet another worldwide conflict, a continents-wide struggle that would prove more terrible and destructive than the first. World War II would provide the impetus for the airplane's next evolutionary leap.

The NACA Langley hangar, circa 1933.

Swords and Plowshares 1940–1957

During the dry season in 1923, as the Curtiss "flying boats" of the forest patrol swooped low over the Canadian timberland, alert to any sign of fire, a 7-year-old boy watched in admiration and envy. Often, he would wave; from his forward perch in a former gun turret an observer returned the greeting. The more the boy saw of the airborne foresters, the more impressed he became. Soon, he began to picture himself as an aviator, in command of powerful aircraft, carrying out important and useful missions. By the time the boy returned several years later to Michigan's Upper Peninsula, the place of his birth, a new goal had crystallized: he would become a test pilot.

Langley test pilots (from left) Mel Gough, Herb Hoover, Jack Reeder, Steve Cavallo and Bill Gray stand in front of a P-47 Thunderbolt fighter in this 1945 photo.

By 1943 the young boy's dream had been realized, for now the man was an NACA test pilot flying out of Langley Field. Today, he was flying a Vought F4U Corsair for the first time. Attached to

NACA research was instrumental in the development of the major experimental aircraft designs of the 1940s and early 1950s, seven of which are shown here at Edwards Air Force Base in California.

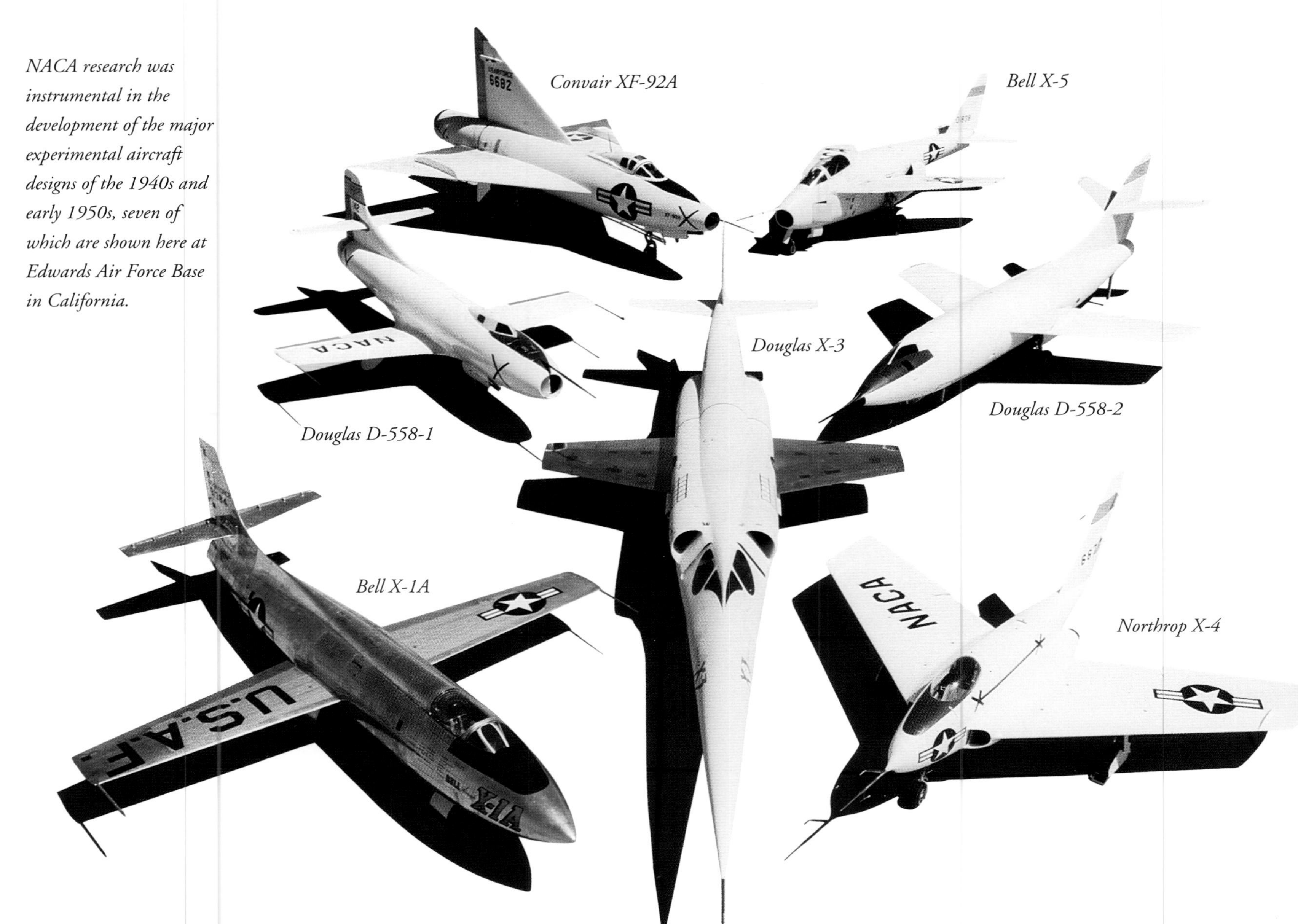

the craft's motor was a hydraulic torquemeter, a device used to monitor and measure engine power. It appeared a routine outing, one of many flight tests conducted at Langley during the war years. Suddenly, at 4,000 feet over the nearby town of Newport News, the pressure line connecting the torquemeter to the engine broke. Almost immediately a thick coat of oil streamed along the airplane fuselage and up over the canopy.

In order to see, the pilot was forced to open the canopy, but in so doing was soon covered in oil himself. His goggles also obscured, the aircraft too slippery for a safe bailout, the pilot decided to return to Langley. As he approached the Field, struggling to see out of one barely open eye, observers said that the plane appeared to be on fire.

Now retired from Langley, John P. Reeder can smile at the recollection. The former test pilot survived his brush with catastrophe, flaring his F4U to a safe landing just past the tail of a parked B-24. "I wasn't jittery or shocked after it was all over. I was too busy thinking of how to get out of the situation," Reeder recalls. "I really did enjoy my flying, even though I had to handle some pretty wild beasts. Many were unstable—they'd fly sideways, speeds would vary. We'd fly because we were trying to find something wind-tunnel tests hadn't shown. You can't get handling characteristics from a wind tunnel."

In a way that went beyond symbolism, the test pilot was the bridge between two ages. If the old aeronautical age was epitomized by the self-sufficient, ingenious individualism of the Wright brothers, then the new aerospace age would be characterized by coordinated group effort between teams of researchers, between humans and new generations of powerful machines. The challenges posed by flight were becoming more and more complex; no one individual could go it alone. Humankind was beginning to reach beyond the usual boundaries, beyond the speed of sound, beyond the lower reaches of the atmosphere, even beyond the familiar grasp of Earth's gravity. Highly trained, disciplined, in excellent physical shape, the test pilot would be the point of the human exploratory spear.

Vought F4U Corsair

The technological explosion that brought the word "aerospace" into use was fueled by the outbreak of a second world war. The requirements of that widespread, mechanized war pushed technology to the point where rapid scientific advance came to be taken for granted. Radar, jet aircraft, the atomic bomb, intercontinental ballistic missiles, rockets, computers, communications satellites, spacecraft: these were but a few of the offspring spawned by a conflict that spanned oceans and continents.

For Langley, the Second World War proved a watershed in several ways. First, the Laboratory's total working staff (professional and non-professional) increased by more than 240 percent, from 940 at the end of 1941 to 3,220 by the end of 1945. The pace of technology development accelerated; airplanes were flying faster, higher and farther. In addition, Langley did not remain NACA's sole research facility. In the late 1930s two additional aeronautical research labs were authorized, Ames Aeronautical Laboratory in Sunnyvale, California, and the Lewis Flight Propulsion Laboratory outside of Cleveland, Ohio. By 1940, Langley had two junior siblings with which it shared talent and accumulated experience. While friendly collaboration between the three was the norm, there was also rivalry—tolerated, as *Engineer In Charge* author James Hansen writes, "only to the extent that duplication, competition and cross-fertilization were productive."

Pearl I. Young, the NACA's first female professional, at work in the instrument research laboratory circa 1929.

Technicians are pictured installing flaps and wiring on a flying-boat model, circa 1944.

War would bring societal change, not the least of which was the increased presence of women in Langley's professional work force. Proportionally speaking, the female presence in engineering science was slight, even though many of Langley's human "computers"—those who assisted engineers by performing mathematical computations by hand on bulky adding-machine-like devices—were women. (This was a fact that pleased some of the Laboratory's male staff who, quite literally, married their computers.)

Langley's human computers at work in 1947.

With the large wartime increase in staff levels, Langley lost some of its clubby, brain-trust feel. Although the Laboratory continued to seek out exceptional engineers and researchers, some of more modest abilities came to work in Hampton. Nevertheless, the World War II years and the period following were among Langley's most exciting and productive. In a world where one "hot" war had ended and a "cold" war was about to begin, the question was how to abide by the biblical edict to beat swords into plowshares: that is, how best to adapt machines of war to serve mostly, although not exclusively, peaceful purposes. The answer, at least for those in the aeronautical community, was a full-scale sprint toward jet propulsion and its affiliated technology. Close on the jet's heels were satellite- and human-carrying rockets.

The Slippery Slope

Even as the bloodiest war in human history raged, NACA Langley continued its work in the relative calm of Hampton. During World War II, the Laboratory temporarily shelved basic research and concentrated on short-term goals, namely, the rapid betterment of existing military-aircraft design. There was little doubt that improvements were essential. The Germans and Japanese had produced several superb

Langley's drag-cleanup studies of the Brewster Buffalo experimental fighter in 1938 were so productive that both the U.S. Army and Navy sent most of their World War II prototype and production aircraft to the Laboratory for similar examination.

aircraft. In particular, Axis fighters threatened to dominate in aerial combat. If the United States and her allies were to emerge victorious, then Allied fighters had to be equally agile and fast.

A Curtiss SB2C dive bomber is prepared for tests in the Full-Scale Tunnel.

By the late 1930s, Langley had been called upon by aircraft companies and the military to examine the latest American military-airplane prototypes. Over the next several years lives would literally depend on how fast airplanes flew and how efficiently they used fuel. The primary means used to accomplish this was to streamline as much as possible the entire aircraft surface. Drag reduction, or "cleanup," improved considerably military-airplane performance.

The Brewster XF2A "Buffalo" was Langley's first full-fledged effort at drag cleanup. The craft was brought to the Laboratory's Full-Scale Tunnel in April 1938 for study. At the end of 5 days of intensive tunnel testing, Langley researchers had suggested ways to eliminate drag caused by the craft's landing gear, exhaust stacks, machine-gun installation and gun sight. The proposed changes raised the Brewster's speed to 281 mph, from the original 250. The 31-mph boost amounted to more than a 10 percent increase in performance.

"We almost took that airplane apart," recalls Herbert A. Wilson, who came to Langley in 1937 and who retired as the Center's assistant director for Space in 1972. "One of the first things we found—and it was very important in World War II—was that the initial cowling design didn't pay too much attention to the air flowing through it. Reducing the amount of air

A Lockheed YP-38—a prototype of the famous Lightning series—undergoing wing-modification investigation in the Full-Scale Tunnel in 1941.

flowing into the engine and redirecting it as it flowed out amounted to a significant increase in performance. For one, it cut down on the amount of fuel needed for a given range. For another, it increased the maximum speed."

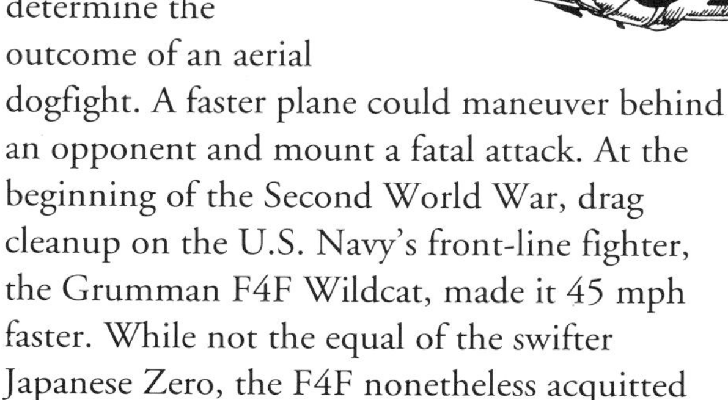
Grumman F4F-4 Wildcat

Extra speed, even as little as a 15-mph edge, could determine the outcome of an aerial dogfight. A faster plane could maneuver behind an opponent and mount a fatal attack. At the beginning of the Second World War, drag cleanup on the U.S. Navy's front-line fighter, the Grumman F4F Wildcat, made it 45 mph faster. While not the equal of the swifter Japanese Zero, the F4F nonetheless acquitted itself well in combat, attaining speeds of up to 320 mph. The F4F's successor, the F6F Hellcat, was faster and more maneuverable, able to reach a maximum speed of 375 mph. The Hellcat, which destroyed nearly 5,000 enemy planes in aerial engagements, is considered by many aviation historians to have been among the best aircraft-carrier-based planes flown by the Navy during World War II.

The Laboratory's meticulous design-analysis efforts spared no detail. Researchers devised one program in which simulated rivets of varying sizes were mounted, row by row, from the nose backwards, on a series of smooth wings. At each stage the drag caused by the rivets was carefully measured. The results indicated the precise amount of drag induced by a given rivet's size and location. Langley's tests indicated that, in order for aircraft to efficiently attain maximum speed, flush—nonprojecting—rivets should be routinely used. Flush riveting soon became standard on both military and commercial airplanes.

Similar Langley programs focused on other aircraft components. Modifications were made almost piece by piece. "In the end you knew what percentage of drag was associated with each piece [of the airplane]," says Laurence K. Loftin, Jr., who arrived at Langley in 1944 and retired in 1973 as director for Aeronautics. "The idea was to make airplanes faster. And we did."

The original NACA cowling underwent substantial improvement, as contours were modified to retain low-drag characteristics at speeds approaching 500 mph. Laboratory researchers examined and solved problems with landing gear not properly retracted or fairings that didn't properly cover the retracted gear. Some manufacturers failed to correctly smooth the area where the wings joined the airplane fuselage or created poor angles between

The effectiveness of a high-speed cowling, installed on this model of a Vought Corsair F4U-1, was examined in the 8-Foot High-Speed Tunnel during April 1943.

windshield, canopy and fuselage, all oversights that resulted in higher-than-necessary drag. These problem areas, too, were investigated at Langley and solutions were proposed. Researchers also worked to identify basic design flaws, such as the ones that caused a buildup of carbon monoxide in the cockpits of certain U.S. Navy fighters. A poor canopy and fuselage layout allowed the odorless but deadly gas, a by-product of engine combustion, to pass into the pilot's compartment.

When late in 1941 the Lockheed P-38 Lightning began to experience problems recovering from high-speed dives, Langley was asked for help. Three months later, after an extensive series of tests in the 8-Foot High-Speed Tunnel, Laboratory researchers had devised a dive-recovery flap. Installed on the lower surface of an aircraft's wing, near the leading edge, the wedge-shaped device created just enough lift so that pilots could regain control of their craft. Although a significant wartime contribution in its own right, the flap would also prove of use during Langley's determined research attack on the "transonic" flight regime, that region where speed increases from just under to just over the speed of sound and where large changes in aerodynamic forces occur. Faster-than-sound flight was only to be achieved after the war, but World War II pilots were already beginning to experience problems relating to high aircraft speeds.

Dozens of aircraft passed through the Laboratory on their way to a better wartime design and thence to combat duty. During one month alone, in July of 1944, 36 U.S. Army and Navy planes were evaluated in detailed studies of stability, control and performance. Langley tested 137 different airplane types between 1941 and 1945, either in wind tunnels or in flight, including virtually all types that actually saw combat service.

By the late 1930s, a Langley team led by Eastman N. Jacobs had developed a series of airfoils designed to delay the onset of aerodynamic turbulence. As airplanes fly through the atmosphere, air flows over the surfaces of wings in a series of layers. The layers closest to the wing's leading edge are smooth or, in the parlance of aerodynamicists, "laminar." But at some point on the wing, and depending on design, the smooth flow becomes turbulent as the air layers bunch up and mix together. If it were possible to delay the onset of the separation of those air layers and the drag that resulted, then there would be big payoffs in an airplane's speed, its cruising range, its

The dive-recovery flaps on this P-47 Thunderbolt are barely visible underneath the wings.

use of fuel, or combinations thereof.

In Langley's wind tunnels, the so-called laminar-flow airfoils performed well. The air flowing over model wing sections—kept smooth and clean by constant attention—did indeed exhibit laminar-flow properties over a relatively large surface. Test flights, though, were another matter, revealing that true laminar flow was extremely difficult to achieve. Part of the problem was keeping wing surfaces clean of debris, a next-to-impossible task given the way planes were manufactured—there were plenty of small crevices where dirt could accumulate—and less-than-ideal operating conditions—mechanics soiled the aircraft as they maintained or repaired it, and dead insects fouled surfaces on landings and takeoffs.

The project was nevertheless trumpeted as a technical triumph by NACA officials ever eager to impress a tightfisted U.S. Congress with NACA research prowess.

Although the project was oversold, Langley's laminar-flow efforts did lead to an airfoil-shape series that was first employed on North American Aviation's P-51 Mustang, which first flew in 1941. The Mustang went on to become a highly effective escort fighter for long-range bombing missions in World War II. In fact, this later-named "low-drag series" was so successful in improving aircraft performance, especially at high subsonic speeds, that its airfoil shapes continue to be used by airplane designers to this day.

Some observers have criticized the NACA's wartime efforts as too shortsighted. In this view, while Langley was solving a host of specific war-related problems, equally important fundamental research—notably into jet propulsion and rocketry—went undone. Failure to

In this 1950 aerial view of Langley, the original East Area is at the bottom of the picture, along the Back River. The West Area, developed early in World War II, is at the top.

A 1947 bird's-eye view of the East Area hangar complex. In the foreground sits a Douglas C-54 transport, flanked by two Boeing B-29 bombers; next to the river is the Full-Scale Tunnel, adjacent to the 19-Foot Pressure and 20-Foot Spin Tunnels.

In 1946 Langley equipped a North American P-51B Mustang with wing gloves for an investigation of low-drag performance in flight.

pursue fundamental research in these areas, some individuals maintain, hindered the nation's progress in the new field of astronautics. Defenders counter that Langley's wartime focus on improving subsonic military airplanes was proper, indeed essential. "The thought on the part of military planners was to stick with one thing," Herbert Wilson says. "It's for that reason that we were somewhat behind the Germans in rocketry. It wasn't for lack of imagination on our parts. If we had divided our efforts we might have compromised our ability to win."

As in any victory, however, the attention of the victor must inevitably turn to new conquests. In Langley's case, it was that of passing through an invisible and difficult-to-understand barrier.

A Need for Speed

Flying as fast as 100 mph seemed impossible to the pioneer aviators of 1910. Thirty-five years later, 100 mph appeared ridiculously slow for everything but recreational flying. During the war years the need for speed was indeed a real one, as pilots sought to outmaneuver and outfight their opponents. Even after—especially after—the cessation of hostilities, fascination with faster and more powerful planes took hold and would not let go.

By the end of the Second World War, the Germans and the British had a handful of operational jet fighters, and the Americans had begun to fly jet prototypes. In the 10 years between 1948 and 1957, the speed of service fighters in the U.S. Air Force and Navy virtually doubled, from 670 to 1,200 mph. A speed faster than that of sound—760 mph at sea level at moderate temperatures, 660 mph at altitudes above 36,000 feet, where temperatures average -60 degrees Fahrenheit—would be attained by Captain Charles E. "Chuck" Yeager on October 14, 1947, in the rocket-propelled X-1. By November of 1950, the first jet-to-jet dogfight took place over Korea. In May of 1952, the first regularly scheduled jet passenger service began with the flight of the British-built Comet. By 1954, a prototype of the Boeing 707 had taken to the air; in that same year, Pan American World Airways ordered 45 jet transports. By the late 1950s, jet transports were routinely flying across the continental United States and to Europe.

The advent of the jet and its penetration into military and commercial spheres would change habits and lives, make a global economy possible in succeeding decades and spur further aeronautical advances. Although high-speed flight research had been conducted at Langley since the late 1920s, there were enormous technical challenges in making such speeds practical. But the concentrated energies of Langley researchers would, in the 1940s and 1950s, lead to a more complete understanding of high-speed flight. Results of such work at Langley and elsewhere would enable, first, military jets and, later, commercial aircraft to fly at speeds only dreamed of in prior years. In the 1930s Laboratory staff were the first to develop highly efficient airfoil shapes used in the design of high-speed propellers; in the 1940s they were among the first to explore practical methods of traveling beyond the apparent "sound barrier."

America's first jet airplane, the Bell P-59, undergoing drag-cleanup tests in the Full-Scale Tunnel in May 1944.

It was in 1938 that British aerodynamicist W. F. Hilton first used the phrase "sound barrier" in remarks made to a reporter. Hilton said that an airplane wing's "resistance" to high speeds "shoots up like a barrier" the closer to the speed of sound an airplane travels. (High flight speeds are often expressed in Mach-number multiples, as a tribute to Austrian physicist Ernst Mach, famed for his exploration into the physics of sound. Mach 2, for example, is twice the speed of sound, or 1,320 mph at 36,000 feet or higher.) But to fly at "super" sonic speeds would present vexing challenges, ones that worried designers and engineers alike. Could aircraft be controlled at such high speeds? Would structures survive higher stresses and temperatures? Was supersonic flight at all practical?

Langley researcher Robert T. Jones was the first American aerodynamicist to identify the importance of sweptback wings in efficiently achieving and maintaining supersonic flight.

NACA Langley used this modified B-29 Superfortress to drop weighted test models from altitudes of 35,000 to 40,000 feet to study aerodynamic forces at transonic and supersonic speeds.

"A lot of people thought for years that it was impossible to fly through this sound barrier," observes former Director for Aeronautics Laurence Loftin. "The thought was, if you bump into this invisible wall in the sky your aircraft would go to pieces. Indeed, there was some experimental evidence that this was the case. A number of pilots were killed trying."

The chief difficulty was that of compressibility effects. The nearer to sonic speeds became, the more aircraft were subject to a sharp increase in drag and a dramatic decrease in lift. In such extreme circumstances—extreme, that is, compared with "average" subsonic flight—control surfaces of traditional propeller-driven planes didn't respond well, if at all. Some pilots in World War II, finding themselves in near-supersonic, fatal dives, literally bent their control sticks in a vain attempt to pull up in level flight. Others—the majority—managed to pull their planes up at lower altitudes.

In 1945 Langley staffer Robert T. Jones was the first American aerodynamicist to realize that the angle at which airplane wings were placed in relation to oncoming air—their "sweep"—would make a critical difference in achieving and maintaining supersonic flight. Jones' calculations indicated that, at faster-than-sound speeds, the air flowing over a thin sweptback wing would actually be subsonic, thereby delaying or preventing compressibility effects. Swept wings were a significant aeronautical advance and eventually wound up on nearly every high-performance military airplane. After 1950, wing sweep was also incorporated in the design of commercial aircraft in order to increase aerodynamic efficiency at high subsonic cruise speeds, between Mach 0.8 and 0.85.

For all the desire to get aircraft safely through the figurative barrier of sound, the obstacles were formidable. Particularly vexing for wind-tunnel researchers was their inability to precisely measure the transonic transformation from pure subsonic to pure supersonic flow. To better understand the nature of the transition, in the mid-'40s researchers employed several methods to collect accurate data. One of the most productive involved dropping from high-flying aircraft bomb-like devices containing electronic gear. These "drop bodies" were then tracked by radar. Information on airspeed, readings of atmospheric pressure, temperature and the like was relayed via a small radio transmitter placed inside the drop body. Many NACA engineers considered these data reliable enough to estimate the drag and power requirements of a future transonic airplane; indeed, test results were incorporated into the design of the sound-barrier-breaking X-1 aircraft.

Another, earlier method was termed "wing-flow technique" and entailed the mounting of a small model wing perpendicular to the wing of a P-51 Mustang. The Mustang took off, flew to altitude and initiated a series of steep dives. For brief periods during the dives the air would flow supersonically over the model. A small balance mechanism fitted within the P-51's gun compartment and tiny instruments built into the mount of the model recorded the resulting forces and airflow angles.

Still another means of transonic investigation included test runs in the Annular Transonic Tunnel, which, in essence, was a whirling arm to which a model was attached. There was some question as to the accuracy of the Annular Tunnel data, but it did provide information on airfoil pressure distributions at speeds of Mach 1—the first ever thus collected. In addition, a "transonic bump" was installed on the floor of the 7 × 10-Foot Tunnel.

As air flowed over the bump, to which was attached a small model, the airflow accelerated to transonic velocities even though the main flow remained subsonic.

However ingenious these attempts were, the fact remained that larger-scale wind-tunnel testing was the preferred method of evaluating the transonic regime. Experiments could be made upon bigger (even full-scale) models, more accurate information collected and then repeated to verify initial results. But researchers attempting to increase wind-tunnel speeds encountered a phenomenon known as

An illustration of the wing-flow method—mounting small models on airplane wings—used in transonic flight research.

Langley's development of slotted walls for wind tunnels—here seen installed in the 16-Foot High-Speed Tunnel—permitted, for the first time, a smooth transition from sub- to supersonic airflow. The advance is widely considered a benchmark in aeronautical research.

Navy combat air patrol aircraft model, tested at Langley, shows two extreme positions of variable-sweep wing.

Two models of the Air Force's Convair F-102 sit poised for launch from Langley's Wallops Island facility. The "Coke-bottle" shape of the model on the bottom follows the area rule.

"choking." As airflows increased to near the speed of sound, shock-wave interference patterns would form, thereby skewing the results of tests. Fortunately, a Langley team led by John Stack and Ray H. Wright discovered that the placement of slots along wind-tunnel walls reduced or eliminated the interference. The development of this "slotted-throat" wind tunnel was an important advance. Writing with Richard Corliss in *Wind Tunnels of NASA*, Don Baals elaborates upon the significance of the find:

> Nowhere in the annals of aeronautical history can one find a more convincing argument supporting fundamental research than in the success story of the slotted-wall tunnel. [It was] a breakthrough idea… a long-sought technical prize [which] …ultimately led directly to the discovery of the famous Area Rule, which in turn spawned a whole new generation of aircraft. So important was the slotted wall in aviation research that in 1951 John Stack and his associates at Langley received the coveted Collier Trophy for their work.

Early in 1947 promising test runs of the slotted-throat concept were made in a 12-inch model tunnel. By the end of 1950 the concept was applied to full-scale facilities, as slots were installed in both the 8-Foot and 16-Foot High-Speed Tunnels. Results were, to say the least, encouraging. Initially classified, Langley's slotted-throat breakthrough was made public in the early 1950s, and transonic researchers worldwide quickly altered their wind tunnels to incorporate the modification.

Unique transonic-design, aerodynamic and propulsion research conducted at Langley was in part responsible for that October day in 1947 when Chuck Yeager briefly broke through the "barrier" of sound in the rocket-powered X-1, the first of a series of high-speed research aircraft. (The 1947 Collier Trophy went to Yeager, Langley's John Stack and Bell Aircraft Corporation president Lawrence Bell in recognition of their research accomplishments in faster-than-sound flight.) But Langley had not yet finished its work. There remained a good deal to learn about getting to supersonic flight; breaking the barrier didn't mean that aircraft were automatically and immediately able to fly supersonically. The sound barrier was broken by brute force, with rockets, but no aircraft manufacturer in its right mind was going to build commercial or military planes that used high-cost, limited-range rockets. Other means would have to be found.

In transonic studies done in the newly modified 8-Foot High-Speed Tunnel, it became apparent that, as an airplane approached the speed of sound, two different shock waves built up: one on the fuselage and one on the trailing edge of the wing. It didn't appear that conventional designs—the most common was a thick, bullet-like, pointed-nose shape with wings and a tail—would allow an airplane to crack Mach 1. These results were of particular concern to one aircraft manufacturer, Convair, which was building the country's first supersonic fighter-interceptor, the YF-102. Enter Langley researcher Richard Whitcomb with the solution,

The bullet-shaped Bell X-1, piloted by Air Force Captain Charles E. "Chuck" Yeager, broke the "sound barrier" on October 14, 1947.

an idea that thereafter became known as the area rule.

"We had a transonic wind tunnel and a big drag problem. I was going to use the tunnel to find out what happens to the airflow as it goes around an airplane near or at the speed of sound," says Whitcomb, who began to work at Langley in 1943 and who, in 1980, retired as head of the Center's Transonic Aerodynamics Branch. "In 1950 there were no theories to explain it, and yet we had to figure out what was going on. So I collected data and sat there with my feet propped up on my desk and said, 'What the hell's going on?' The shock patterns around the plane weren't what you'd expect. There was a shock wave on the wing that came all the way across and hit the fuselage. I had [German aerodynamicist] Adolf Busemann's data in front of me and it suddenly came together, just like the light bulb that lights up in a comic strip.

"The basic idea was to consider the airplane as a whole, a total entity. It can't be looked at as a collection of separate components. That's what the shock wave was telling us. You had to include the whole area. That's where the words 'area rule' came from."

Whitcomb visualized making more room for the air streaming along the fuselage and wings of an airplane about to go supersonic. The shock waves observed in wind-tunnel studies were caused by a violent intersection of air and plane. Whitcomb's flash of inspiration: pinch the waist of the fuselage in the region of the wing. Air would still be displaced, but not nearly to the extent it otherwise would be. It was a brilliant insight. Soon, aircraft designers would be talking of the "Coke-bottle effect," referring to the visual consequence of the area rule's application.

Photographed in the 8-Foot High-Speed Tunnel in April 1955, Richard Whitcomb examines a model designed in accordance with his transonic area rule.

Because of its military significance, the area rule proved a hot national-security potato, and so was kept secret until September of 1955,

Second in a series of successor research craft to the X-1, the Bell X-1B was photographed at Langley during instrumentation tests.

BELL
X-1B

when its revelation triggered a blizzard of publicity. The National Aeronautic Association awarded Whitcomb the 1955 Collier Trophy, saying, "Whitcomb's area rule is a powerful, simple, and useful method of reducing greatly the sharp increase in wing drag heretofore associated with transonic flight … [and is being used] in the design of all transonic and supersonic aircraft in the United States."

By any standard, the period 1940 to 1955 had been an extraordinary period for aeronautics. Within a 15-year stretch, Langley researchers had a hand in raising aircraft speeds from hundreds to thousands of miles per hour. Emerging from Langley-led research was a historic series of high-speed aircraft, beginning with the sound-barrier-breaking X-1 and continuing with the X-2, X-3, X-4 and X-5. Each aircraft was designed to study different but interrelated aspects of high-speed flight. But the Laboratory's accomplishment was not simply the straight-line result of wind-tunnel investigations and flight tests under rigorously controlled conditions. Rather, it was the associative power of human intellect and intuition that, combined with an exacting scientific method, enabled fundamental advance.

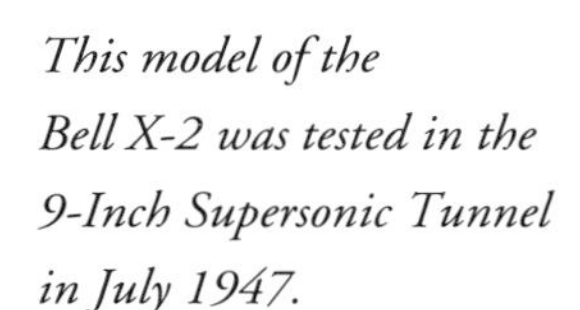

This model of the Bell X-2 was tested in the 9-Inch Supersonic Tunnel in July 1947.

"Both the slotted tunnel and the area rule derived largely from pictures in the mind," writes James Hansen in *Engineer In Charge.* "Achievements by Langley researchers were products of intelligent guesswork, reasoning by intuition, and cut-and-try testing as much as products of numerical systems analysis, parameter variation, or theory."

Langley's study of the supersonic regime was but an introduction to even higher speeds. The Laboratory entered into "hyper" sonic research with the hope of understanding and predicting the flight of planes, rockets and missiles at or above Mach 5. At the time, few realized how close humanity was to the Age of Space.

Faster Than Fast

By late spring of 1944, shortly before D-Day and the Allied invasion of Normandy, it was beginning to occur to even the Nazi High Command that the prognosis for Axis victory was poor. In an attempt to recapture the initiative, the Germans unleashed the first of their secret weapons: the "Velgeltungswaffe Ein"—or, in English, "Vengeance Weapon Number One," the world's first cruise missile. The subsonic V-1 and, later, the supersonic V-2

A Langley model maker examines the molds used to form a model of the Bell X-5, a variable-sweep craft that first flew in June of 1951.

rockets screamed down upon British cities and countryside in what proved to be a vain attempt at intimidation.

One year later, as the "Thousand Year" Reich disintegrated before the relentless Allied onslaught and the advancing armies overran the German rocket-research town of Peenemunde, the true significance of Germany's undeniable technological triumph became chillingly clear. Nazi engineers had intended to design long-range ballistic missiles, two of which—the A-9 and A-10—were planned for the aerial bombardment of the eastern United States. The Allied discovery of the German rocket-research facility had tremendous psychological impact. If the Germans had succeeded with their ambitious undertaking, World War II might well have had a different outcome. The victorious Allied powers realized full well that no spot, however remote, would be safe from military attack if rockets, wedded to warheads of the atomic variety, were only minutes away from delivering their deadly cargo.

Over the next few decades, those countries that could developed their own ballistic missile arsenals to guard against real or perceived threat. The embrace of rocket technology would prove a two-edged sword. On one hand, it would make possible humanity's leap into space. On the other, it would create new weapons of mass destruction, thereby altering the course of world military and political history.

Long before World War II Langley researchers had been aware that jets, missiles or rockets traveling at high-Mach-number speeds would one day be built. But at that time the problems confronting would-be designers were formidable. Hypersonic speeds appeared too much for even the most advanced aerodynamic devices. Rapid passage through the atmosphere generated an enormous amount of frictional heat, heat well beyond the structural tolerance of most metals or metal alloys. But with speeds in Mach multiples a foregone conclusion, new ways to put missiles or proposed hypersonic aircraft together had to be considered.

Langley's first vertical takeoff and landing (VTOL) model.

Research on just how to do so was undertaken in facilities like Langley's 11-Inch Hypersonic Tunnel, which began operations in the fall of 1947 and was the first of its kind in the United States. Built as a pilot model for a larger hypersonic tunnel—the Continuous-Flow Hypersonic Tunnel, itself built 15 years later—the 11-Inch Tunnel operated for 25 years until 1973, when it was dismantled and given to Virginia Polytechnic Institute and State University in Blacksburg, Virginia, for educational uses. In 1951, another of Langley's hypersonic facilities came on-line: the Gas Dynamics Laboratory. There, hot, highly pressurized air released in short bursts from huge storage tanks was funneled to test cells to simulate speeds of up to Mach 8.

Hypersonic research at Langley in the late 1940s and early 1950s focused first on the difficulties long-range missiles would encounter during intercontinental flights. There were many. A successful intercontinental ballistic missile would have to be accelerated to a speed of 15,000 mph at an altitude of 500 miles and then guided to a precise target thousands of miles

A one-tenth-scale model of the X-15 research plane is prepared in Langley's 7 × 10-Foot Wind Tunnel for studies relating to spin characteristics.

away. Sophisticated and reliable propulsion, control and guidance systems were therefore essential, as was the reduction of the missile's structural weight to a minimum. And there was aerodynamic heating, which could cause the missile's nose cone to heat up to tens of thousands of degrees Fahrenheit.

The same problems that confronted missile makers would later face spacecraft designers as they attempted to boost a human cargo safely into orbit and return it just as safely to Earth. Langley's Structures Research Division, which had in the '40s and early '50s concentrated on aircraft flutter and vibration problems, took on the materials question, work that eventually led to successful reentry designs for space capsules.

The aerodynamic heating issue was addressed by former Langley employee H. Julian Allen, who had moved to a new post as chief of High-Speed Research at the NACA Ames Laboratory in California. Allen devised the "blunt-body" concept, which did away with the idea of a sharply pointed nose in favor of a rounded shape. Upon atmospheric reentry, the blunted form caused the formation of a shock wave, which dissipated most—although not all—of the frictional heat into the atmosphere. Missiles and spacecraft could therefore be made, if with some difficulty, to survive a searing return to Earth. The blunt-body approach was subsequently incorporated into the designs of the Mercury, Gemini and Apollo astronaut capsules.

The Langley-led X-15 project, a joint effort undertaken by the NACA and the military, was initiated in 1954 to tie together all of the hypersonic research then underway. North American Aviation pilot (and former NACA test pilot) A. Scott Crossfield was at the controls as the X-15, the world's first hypersonic research airplane, undertook its maiden flight on June 8,

X-15 launch techniques were investigated using one-twentieth-scale models mounted in the 7 × 10-Foot Tunnel.

1959. In investigations intended to gather data on aerothermodynamics, structures, flight controls, and human physiological reactions to high-speed, high-altitude flights, three X-15s flew a total of 199 missions between June 1959 and October 1968. Perhaps most importantly, the X-15 served as the "test bed" for techniques and systems that later would be employed in the development of the Space Shuttle. As author James Hansen writes: "The Shuttle's reentry characteristics—the transition from the reaction controls used in space to aerodynamic controls, the use of high angles of attack to keep the dynamic pressures and the heating problems within bounds, and the need for artificial damping and other automatic stability and control devices to aid the pilot—are similar in all important respects to those of the X-15 conceived at Langley."

Until the first orbital flight of the Space Shuttle *Columbia* in 1981, the X-15 held the altitude and speed records for winged aircraft, with flights as high as 67 miles and a maximum speed of 6.7 times the speed of sound, or 4,518 mph. The X-15 program was, agree the experts, one of the most successful aeronautical research endeavors ever undertaken.

"Some have said that the X-15 was the hyphen in aerospace," says John Becker, retired chief of the High-Speed Aerodynamics Division. "Up until 1952 or '53, there was almost no realization that we were on the verge of the Space Age. Then, suddenly, we realized we had the propulsion to get up to hypersonic speeds and also to get out of the atmosphere—at least for a little while—and out into space. When that began to sink in, it became a very exciting period."

Sputnik Shock

Alone among the four major Allied powers, the United States emerged unscathed from the Second World War. Protected from attack by two vast oceans, the American infrastructure had not suffered the terrible devastation experienced in Europe and Asia. Its industrial base vigorous, America prospered, becoming the world's most powerful nation, the first "superpower" in history. By the time Dwight Eisenhower became the nation's 34th president in January 1953, and despite fears of Communist infiltration or aggression sponsored by the Soviet Union, the fact of America's technological dominance was taken for granted. So it was a profound shock when the Soviet Union beat the United States into space on October 4, 1957, with the launch of the world's first satellite, Sputnik. To add insult to injury, less than a month later, on November 3, the Soviet Union sent into orbit a second Sputnik. Sputnik 2 carried a payload many times heavier than the tiny payload planned for Vanguard, the first American satellite.

Renowned American scientific and technological know-how suddenly seemed second-best, overshadowed by an ascendant Communist space science. The beep-beep-beep of the orbiting Soviet satellite took on ominous overtones and was amplified by national doubt and embarrassment until it reverberated across the political landscape like the characteristic boom produced by an airplane going supersonic. Scarcely a year after the Sputnik scare, the NACA was no more—replaced by another agency, NASA, whose implicit priority was to make America number one in space. It hadn't been too long before, as one observer dryly commented, that the NACA stood "as much chance of injecting itself into space activities in any real way as an icicle had [surviving] in a rocket combustion chamber."

At first, things didn't go all that smoothly as the United States played space catch-up. James Hansen explains:

> ...On the sixth of December [1957], with hundreds of reporters from all over the world watching, the Vanguard rocket rose a mere four feet off its pad at Cape Canaveral, toppled over, and erupted into a sea of flames. The international press dubbed the failed American satellite "Kaputnik" and "Stayputnik." Cynical and embarrassed Americans drank the Sputnik cocktail: two parts vodka, one part sour grapes.

At the United Nations, a Soviet delegate even asked sarcastically if the United States should receive aid as an underdeveloped country. But the ridicule was short-lived. Six weeks later, on January 31, 1958, an Army team headed by former German rocket scientist Wernher von Braun managed a successful launch of the 31-pound Explorer 1. At long last, America was in space.

Nationally, changes in aerospace-related government policy were under way. One of the

biggest came in the changeover from NACA to NASA management. Although certainly not a major change in the eyes of employees—nearly everyone retained the same job and responsibilities—over time, the transformation would prove significant. NASA would undertake projects on a scale unheard of in NACA days. As perceived masters of space technology, the new agency would also be held to standards few (if any) government agencies could easily match. Every NASA success was lauded, every shortcoming mercilessly scrutinized. Whether for good or ill, the NACA had rarely, if ever, been put under such a powerful public microscope.

In Langley's case a more local transformation involved the public perception of the "NACA nut." No longer considered technology-obsessed eccentrics by even the most contrary of Hampton residents, Langley Laboratory's research scientists and engineers were becoming Space Age wizards, valued as interpreters of the obscure runes of spaceflight physics and orbital mechanics.

"Conjure the scene from The Wizard of Oz: the wicked witch flies over the Emerald City spelling out 'Surrender Dorothy,'" James Hansen writes, "and all the terrified citizens rush to the wizard to find out what it means. In an exaggerated way, this gives some idea of how the Sputnik crisis and the resulting American space program triggered the local public's feelings of wonder about, and admiration for, Langley."

In the years to come Langley would live up to that admiration in a big way. As the first home to the U.S. manned space program and the first NASA astronaut training center, Langley Research Center would prove that it could learn as much about the practicalities of spaceflight as it already had about the requirements of aircraft flight.

The X-15 made 199 flights between June 1959 and October 1968. Until the maiden flight of the Space Shuttle Columbia *in 1981, the X-15 held the world altitude and speed records for winged aircraft.*

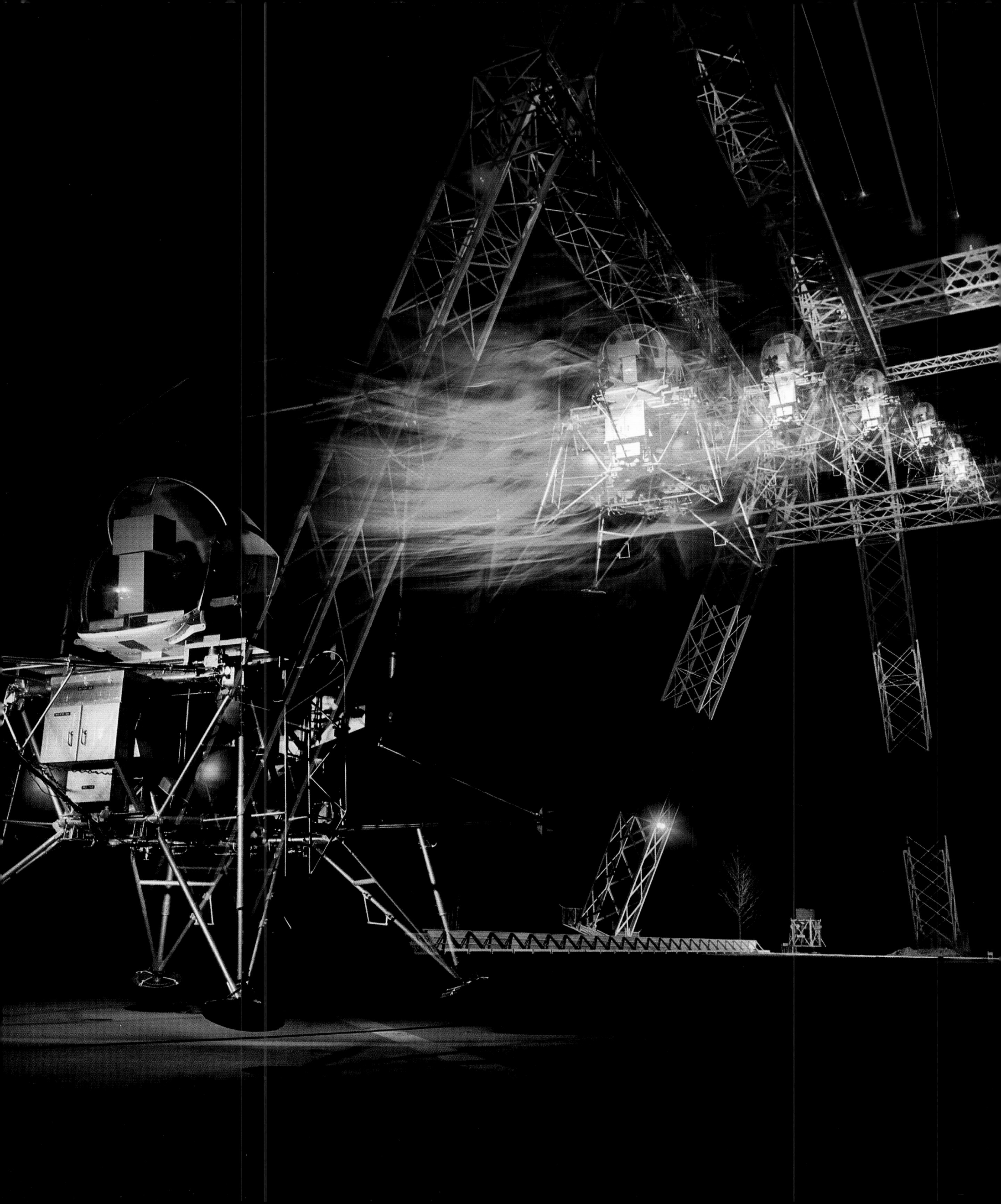

Beyond the Home Planet 1958–1969

Something about a beach soothes the soul. Of comfort is the rhythm of seawater falling upon sand, or the nearly constant wind, or simply the sight of a vast ocean vanishing over the horizon. For the Wright brothers, the appeal of an oceanside site was eminently practical: steady winds could keep research gliders of the sort they designed aloft for quite some time. Landing on sand would also prove gentler on the flimsy structure of the Wright *Flyer*. Too, the Wrights could carry on their work far from the prying eyes of the press.

This multiple exposure photo shows a simulated Moon landing at Langley's Lunar Landing Research Facility. Landing simulations were conducted at night to better simulate the dark Moon "sky."

For a later generation engaged in rocket research, surfside was also the place to be. The secluded Wallops Island range where Langley began testing rocket models in the mid-1940s suited NACA officials just fine, especially since, as part of its overall program, the Laboratory was providing research assistance to the military for a highly classified guided-missile program. In addition, working on the

Liftoff of the 363-foot-tall Saturn V rocket at 9:32 a.m. EDT on July 16, 1969. Aboard the Apollo 11 spacecraft were astronauts Neil Armstrong, Michael Collins and Edwin "Buzz" Aldrin.

island kept inherently dangerous devices away from population centers. In the event of explosion or in-flight destruction, it was far better to have a rocket pulverize over the ocean than over a city. Not that Langley researchers wanted to see their work go up in smoke. On the contrary, successful research-rocket firings from Wallops would furnish much useful information, information that in time would prove invaluable in the American exploration of the high frontier of space.

By 1944, small teams of Langley's Wallops Island researchers were launching rocket models that weighed about 40 pounds. Instruments placed inside relayed information via radio signals to observers on the ground. Although the results helped to further the U.S. Army's ballistic missile experiments, NACA researchers were keenly interested in defining the best airplane wing-and-fuselage configuration and control systems to fly in and through the transonic range. Rocket-model tests helped to improve upon high-speed-research methods and devices. Langley's scientists and engineers developed new ways of measuring, transmitting and recording accurate data even as their small rockets changed speed, altitude and attitude in a matter of a few seconds.

For the individuals working at Wallops in the 1940s and 1950s, Eastern Shore isolation

Takeoff of a five-stage missile-research rocket from Wallops Island in 1957. The first two stages propelled the model to about 100,000 feet; the last three stages were fired on a descending path to simulate the reentry conditions of ballistic missiles.

created a sense of fellowship, in part because of the rugged surroundings. The island was difficult to reach; once there, researchers could expect to stay as long as 6 months. Housing was primitive, a choice of spartan Quonset huts or, for the adventuresome, tents pitched on the beach. Food was plentiful and good, but entertainment was limited. There was a shortwave radio to listen to, card games to be had after dinner, spirited conversation and the camaraderie of the like-minded. All in all, report former Wallops rocketeers, it was one of the most enjoyable experiences of their lives.

After the Wallops complex was administratively transformed in June of 1946 into a separate Langley division, it began to attract attention from other Laboratory departments because of the sheer number of models sacrificed in the name of science. In the 3-year period 1947–49, more than 380 plunged to a watery grave in the Atlantic Ocean. Langley's wind-tunnel personnel complained that such an expenditure was roughly equivalent to the requirements of 10 major wind tunnels. Wallops rocketeers countered that one single rocket-model test, because it provided important aerodynamic data, was comparable to the dollar-for-dollar return from wind-tunnel research.

Whatever the technical or other merits, those working at Wallops were energized by their labors. "The environment at that time was something. I remember thinking, 'You pay people to do this?'" recounts W. Ray Hook, Langley director for Space. "There was great freedom to make mistakes. People didn't fear trying something new. The attitude was, if you think you can do it, try it. We were flying things on rockets at a good clip fairly early in our careers. And we built nearly everything ourselves. You got your own model, assembled your team,

Aside from native flora, fauna and the Langley rocket-research complex, there was not much on Wallops Island. Pictured here is a 1960 photo of Launch Area Number Three, used principally for Scout rocket firings.

The first Scout prepared for launch at Wallops Island July 1, 1960, and launched the evening of that same day.

lit the fuse and graded your 'paper' in front of God and everybody. It was tremendous sport."

Not every rocket went off according to intent. Some experiments had to be rethought even though the basic premise appeared sound. Once, investigators had to scrap plans to send a pig on a 100,000-foot suborbital flight. Although researchers had gone to the trouble of designing a special couch for their would-be porcine passenger, it was determined that pigs can die if they lie on their backs for too long. But an animal finally did make it into space from Wallops, on December 4, 1959, when a successful suborbital test of the Mercury capsule boosted Sam, a rhesus monkey, to an altitude of about 53 miles.

Metallic ears pointed to the heavens, this radio-tracking device kept tabs on Wallops rocket firings.

One important project that was initiated in the late 1950s at Wallops was the Solid Controlled Orbital Utility Test Program—otherwise known as Scout. The program officially began in 1957 with the stated intent of building an inexpensive sounding rocket to carry small research payloads to high altitudes. In May 1958, those goals were further refined: Scout would be a four-stage solid-fuel booster capable of putting a 150-pound satellite into an orbit 500 miles above the Earth's surface. On February 16, 1961, Scout successfully boosted into orbit the Explorer 9 satellite, a 12-foot sphere designed for atmospheric-density measurements. Scout thus became the first solid-rocket booster to orbit a payload, and the first vehicle to do so from Wallops Island.

Scout would eventually assist the Mercury, Gemini and Apollo programs by testing reentry materials, evaluating methods of protecting spacecraft from micrometeoroids and examining ways of overcoming radio blackouts as a space capsule reentered the atmosphere. The Department of Defense used Scout to launch the U.S. Navy's highly successful Transit navigation satellites, satellites that pass 600 miles overhead every 80 minutes broadcasting positioning information used by warships, fishing vessels and pleasure craft. For the Air Force, Scout launched in-space targets that were used to test anti-satellite weapons fired from F-15 fighters. Scout scientific payloads also examined how water vapor and other aerosols have affected the Earth's atmosphere, mapped Earth's magnetic field, and made the first observations of a suspected black hole at the center of a collapsed star.

"I don't think there's ever been another project where government and contractor personnel worked together as closely as they did on Scout," says former Scout Project Manager Roland English. "Partly, I guess, it was the nature of the program. The goal we had, the job we were charged to do, [was make] an inexpensive rocket that could be used by a lot of people. It was a goal you could put your heart into."

Designing, building and flying rockets was—and is—not an easy endeavor. As in any complicated undertaking, perseverance can make the difference between success and failure. Langley's rocket researchers kept at it and in the process accumulated invaluable experience that could not be had in any but the school of hard knocks. The skills of Wallops' rocketeers would be put to a bigger test as the United States took its first steps across the borders of the space frontier.

The Mercury space capsule undergoing tests in the Full-Scale Tunnel, January 1959.

To the Moon by Noon?

On July 29, 1958, President Dwight Eisenhower signed legislation that would spell the end of one federal agency and mark the beginning of another. In remarks made at the signing, Eisenhower said that "the present National Advisory Committee for Aeronautics with its large and competent staff and well-equipped laboratories will provide the nucleus for NASA .... The coordination of space-exploration responsibilities with NACA's traditional aeronautical research functions is a natural evolution ...." That evolution was finalized on October 1, 1958, when the NACA officially became the National Aeronautics and Space Administration.

The substitution of the "C" in NACA with the "S" of NASA (or, as some pundits suggested, the replacement of a cents sign with a dollar sign, referring to the higher cost of bigger projects) did not at first seem likely to cause much of an uproar at Langley. After all, those who left work on Tuesday evening, September 30, 1958, as NACA employees were the same people who would come to work as NASA employees Wednesday morning, October 1. But the transformation was unsettling, at least in a long-term sense. For Langley, it spelled the true end—the first phase of which was the large-scale expansion during World War II—of the small-scale, tightly knit brain trust that had concentrated on specific aeronautical problems since the Laboratory's formation in 1917. There was also a name change: in the case of NASA's firstborn, to the NASA Langley Research Center.

The degree of project difficulty would increase as well. The requirements of space

A model of the Mercury capsule undergoes flotation tests.

Technicians prepare a prototype of the Mercury space capsule, 1959.

travel, manned and unmanned, presented unprecedented challenge. The ranks of middle-level management would swell, aided by the need to organize and carry out large-scale programs. As NASA grew into a much larger organization, scientific and technical proficiency, although still central to the new organization's mission, were not quite as lionized as in the days of the NACA. It was a matter of degree: large institutions tend to reward bureaucratic and political skills. As the NACA gave way to NASA, the engineer would gradually relinquish his traditional role as final decision maker.

In the span of a few years, NASA's entire effective work force would balloon manyfold to include thousands of outside contractors hired to assist in research and to build the complex machines that would chart and travel the Solar System. But perhaps the most significant effect of the NACA-to-NASA transition, seen most clearly in the new agency's large-scale space effort, was on the public imagination. Generally speaking, the public idolized NASA, seeing its scientists and engineers as heroes for a new age, high-tech warriors doing great deeds. Indeed, the effort was heroic, but in ways different than those popularly perceived. But such distinctions were obscured by popular enthusiasms. NASA became the one government agency that could do little wrong, even as its growing pains were obscured by largely favorable publicity and the race to put Americans into space.

Many of Langley's old-school aeronautical engineers, enthusiastic about all things related to flight, were dismayed by the new-found dedication to space. Some would permanently opt for retirement or seek employment in the private sector. Others stayed, but felt that the so-called "Space Race" was nothing more than geopolitical posturing, an expensive, unnecessary boondoggle (one prominent Langley staffer was contemptuous of what he called NASA's "to-the-Moon-by-noon" philosophy). Still others gave newborn NASA its grudging due, but more out of loyalty to the NACA's technological track record. In any event, when President John F. Kennedy declared in a May 26, 1961, speech before Congress that before the decade was out, Americans would land on the Moon, there didn't seem to be a shortage of doubters.

"Two years after the Apollo program was announced, in 1963, I had lunch with two Langley division chiefs," John Becker relates. "They said that Apollo was the most dishonest thing to ever happen in the aerospace industry. They said it was crazy to embark upon a project we know we can't do. I sat there and listened to a long litany of problems. But I was thinking, 'Most of us are engineers trained in the old-fashioned way. We have a lot of new things to learn.'"

The Little Joe launch vehicle being readied for a test launch from Wallops in January 1960...

New things learned would blur an already fuzzy line between matters relating to airplanes and those regarding space travel. There were various degrees of technical or administrative separation between the two areas, but often the very people working on spacecraft had wrestled with the transonic problem, or fretted over issues regarding aircraft instrumentation, or were laboring to improve an airplane's structural integrity. In practical terms, this meant that most Langley engineers would move with ease from working on aeronautics problems one day to addressing space-travel difficulties the next.

...and ascending skyward on a plume of exhaust.

Confident in their own abilities, proud of the NACA's achievements, most NASA researchers were sure they could put American spacecraft into orbit. But they were used to relatively small-scale endeavors. Could NASA carry off its expanded mission with the same skill that the NACA had expressed in admittedly more limited arenas? Former NASA engineer Richard E. Horner, in a May 1972 interview, outlined some of the management problems NASA encountered: "The NACA cadre had the typical technical man's disease at the time: the virus of wanting to do too much, the 'reach exceeds my grasp' problem. When I first joined NASA in June of 1959, I was just flabbergasted at the number of programs that were being attempted .... It was very clear to me that either we weren't going to get anything done on schedule, or we were going to have to eliminate an awful lot of things that we were trying to do in the process .... In making the transition [from NACA to NASA] some management mistakes were made. On the other hand, the way the program evolved, they were able to bridge the management-experience gap very successfully."

As part of the Project FIRE study, technicians ready materials to be subjected to high temperatures that will simulate the effects of reentry heating.

An "in-house" researcher-led program at Langley that aimed to put astronauts in space as soon as possible led directly to the formation, in August 1958, of the Space Task Group (STG). Comprised of Langley rocket-research veterans and others from various Langley divisions, as well as personnel from Lewis Research Center in Ohio, the STG was the 36-man nucleus around which ultimately condensed the entire U.S. manned space program. Names well-known to insiders—men such as Max Faget, Robert Gilruth, Caldwell Johnson, Christopher Kraft—were among the handful of leaders responsible for mounting the successful U.S. assault on space. At the time of the STG's formation, most of these individuals were working at Langley. Langley would remain the STG headquarters site until 1962 and the formation of the Johnson Space Center.

Preparing Project FIRE model capsules to be sent aloft on ballistic missiles.

Even before the Space Task Group was formally organized, its Langley members had begun to develop the concept of the "Little Joe" test vehicle, which became the workhorse of the Mercury program. In pre-STG days Center researchers had also demonstrated the feasibility of a manned satellite program, using existing ICBMs as launch vehicles, and originated the contour couch concept, which was adopted for use in all subsequent U.S. space flights. Once it crystallized, the STG began to address additional technical issues, among them proof of the feasibility of a heat-dissipating shield for astronaut-carrying capsules and the development of astronaut "procedure trainers," later called simulators.

A number of Langley-based programs were designed to support the work of the Task Group. One such was Project FIRE (short for Flight Investigation Reentry Environment), which investigated the intense heat (several thousands of degrees Fahrenheit) of atmospheric reentry and its effects on would-be spacecraft materials such as copper, tungsten, Teflon, nylon and fiberglass.

Building test facilities to simulate such extreme heat was no small technical feat, and Langley engineers relied on several different types of technology. One involved the heating, to 4400 degrees Fahrenheit, of a bed of pebbles made from the metallic element zirconium.

Another method created a brief but intense flame from the action of an electric charge upon a compressed test gas. A third involved the launch of multi-stage sounding rockets from Wallops, by which means reentry speeds as high as Mach 26 were attained.

In this same time period, Project RAM (the acronym stood for radio attenuation measurements) focused on how to transmit radio waves through the plasma sheath that formed around reentering spacecraft. Also undertaken was Project Echo, which led to development of the nation's first "passive" communications satellite. Made from aluminized Mylar plastic, the 100-foot-diameter Echo 1 was a giant, automatically inflatable balloon off which radio signals could be bounced. Launched on August 12, 1960, into an equatorial orbit approximately 1000 miles high, Echo 1 could be seen with the naked eye—a graphic reminder of the American effort to effectively compete with the Soviet Union in space.

Of the many notable achievements of the early years of the Space Task Group, one of the most important was the establishment of the Mercury Tracking Network. For the first time, spacecraft and their human operators were to be actively monitored while in orbit. By any

Echo 1 was America's first passive communications satellite, a 100-foot-diameter aluminized Mylar plastic balloon that reflected radio signals beyond Earth's curvature.

standard, it was a gargantuan and unprecedented undertaking. Organized and managed out of Langley, the tracking network's successful implementation underscored that the Center's engineers had what writer Tom Wolfe would later characterize as "the right stuff."

The work of the STG was absolutely essential to the U.S. space effort. The Group later left Langley to found the Johnson Space Center in Houston and to oversee the Gemini and Apollo projects, but its early work in Hampton set the standards by which subsequent U.S. space success was made possible. Heirs to the NACA problem-solving tradition, the Task Group made it clear to anyone who would listen that exploration of space and flights to the Moon were no longer in the realm of science fiction.

"No Albert Einstein was required. Everything we did at the time was doable," says Israel Taback, who, upon his retirement in 1976, was chief engineer on Project Viking, the Mars exploration program. "We understood trajectories. Developing new boosters, new spacecraft, coming up with rendezvous techniques—it was basically an enormous engineering challenge. The only intimidating thing was the size of the job: thousands and thousands of people working all over the country to put two men on the surface of the Moon. Langley was sort of the parent university."

Still Up in the Air

However preoccupied NASA was in the 1960s with space-related matters, at Langley aeronautics research continued. Much had been accomplished in the previous decade, particularly where subsonic flight was concerned. By 1960, atmospheric flight had seemingly matured to the point that only a few major programs—like a supersonic transport—remained to be done. Langley's aeronautics work in the late 1950s and 1960s, then, would concentrate in these yet-to-be-accomplished areas.

Testing the F-111A's variable-sweep wing on a one-tenth-scale model in May 1965.

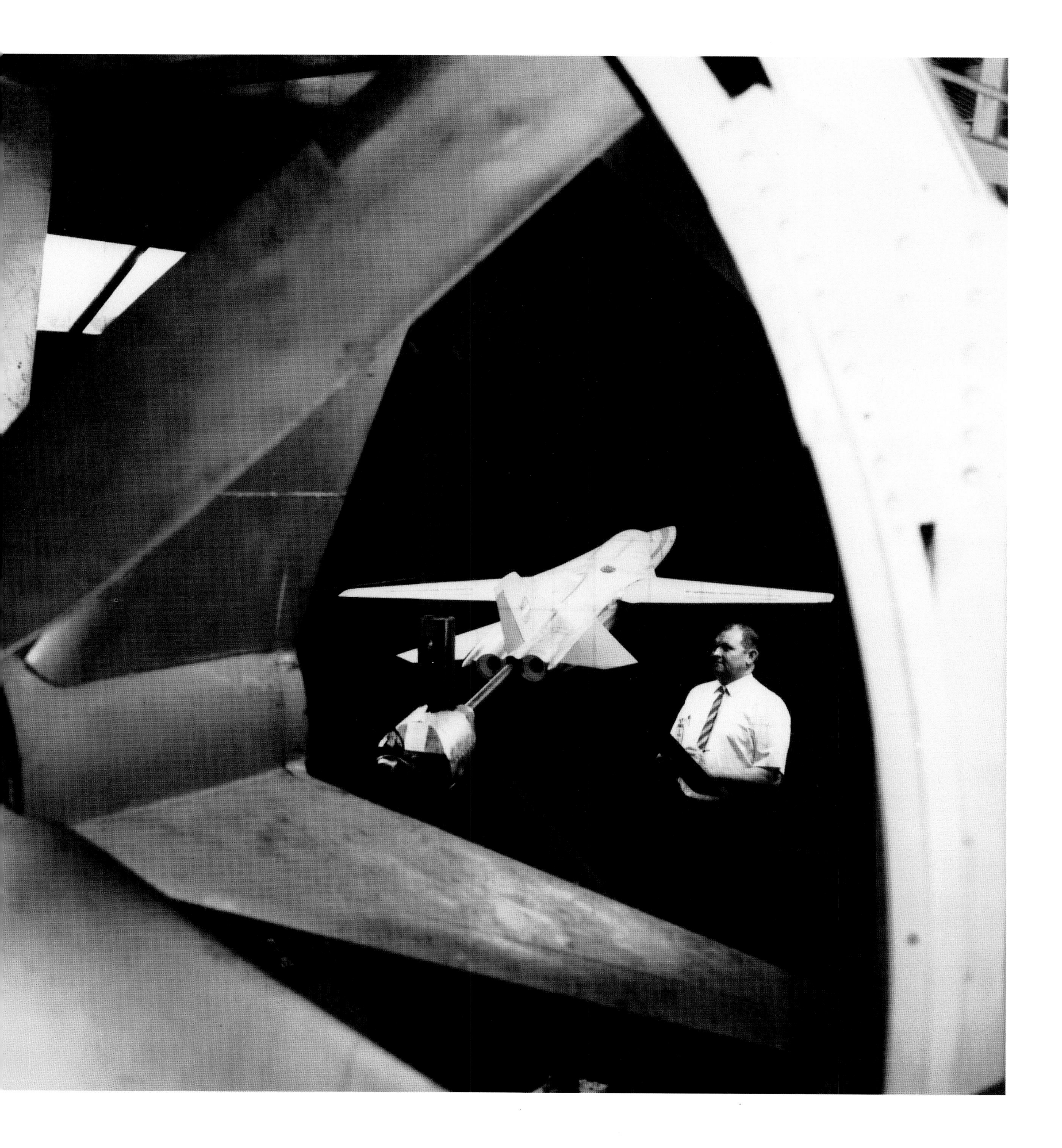

One such area involved the concept of the variable-sweep wing. Simply put, the notion was a variation of swept-wing theory, with this refinement: an airplane's wings could be mechanically adjusted to different sweep angles to conform to either sub- or supersonic flight. At times of takeoff, climb to altitude and landing, the wings ideally would extend almost at right angles to the fuselage, or "near-zero" sweep. When flying faster than the speed of sound, the airplane would resemble more the head of a spear or an arrow, as its wings would be fully swept back.

Although first identified in the early to mid-1950s as a potential means of improving a military airplane's operating efficiency, application proved difficult. Tests made on variable-sweep models indicated that they all suffered from major—and, in the real world, deadly—changes in stability as the wings were rotated through various angles of sweep. Therefore, when Langley-led studies indicated that properly positioning the point at which the wings pivoted would provide the needed stability, it was a notable advance.

To validate the discovery, Langley researchers built four scale models and tested them at transonic speeds in the 8-Foot Transonic Pressure Tunnel. Free-flight model tests were also made. Sweep angles were varied from 25 to 75 degrees and no significant problems, either of stability or control, were observed. One of the most astonishing things about the venture was its speed: Project "Hurry Up" took a little more than 2 weeks from start to finish. As a direct result of the Langley tests, in

A tilt-wing prototype used in vertical takeoff and landing (VTOL) studies. Once aloft, the wing repivoted and the craft would fly horizontally.

1961 the Defense Department gave the go-ahead for production of the nation's first variable-sweep fighter, the F-111. Built by General Dynamics, the F-111 first flew in 1964, entered operational service in 1967 and is still in use today. Variable sweep was subsequently incorporated in the design of many of the United States' fleet of advanced military aircraft.

Also under research scrutiny at Langley in the 1960s were gust alleviation, active boundary-layer control and vertical/short takeoff and landing (V/STOL) systems. Protecting against turbulence caused by wind gusts was of particular concern to the Air Force, which as part of its strategic plan was relying on low-flying bombers in case of war. As a result of tests conducted at Langley in the late '50s and early '60s, structural modifications were made to one model series of the B-52 bomber. (The commercial aircraft industry found little use for the concept.)

An active boundary-layer control system was installed on a prototype Boeing 707-80 airplane in 1964. Large quantities of air were injected parallel to the wing surface and over the leading edge of the craft's flaps to increase the amount of lift at low speeds. The demonstration proved that safe landings could be made with a more efficient use of a plane's power plant and speed-control system.

Building on autogyro research that commenced in the late 1920s and early 1930s, and helicopter research that began during the final years of World War II, in the 1960s Langley undertook to evaluate a variety of V/STOL approaches. V/STOL designs permitted aircraft to rise vertically, helicopter-like, and then fly horizontally. In following years, these designs would be further refined, with the goal of producing a "short-hop" commuter aircraft. A V/STOL craft needs less runway area in which to operate, a fact that may lead to widespread adoption as one way to ease the chronic airport congestion that is predicted to worsen through the 1990s. By the early '90s the concept was still in the research phase and prototypes were continuing to be evaluated.

The duct-fan method of airplane propulsion—which would enable aircraft to take off helicopter-like but fly like an airplane—was tested with models like this one.

By the 1960s, Langley's area rule originator Richard Whitcomb had made another discovery, this one related to the shape of an airplane wing. Whitcomb was looking for ways to delay the onset of the high wing drag caused by localized supersonic flow occurring at high-subsonic speeds. Since the basic airfoil shape was responsible, in his mind's eye Whitcomb visualized an alternative: a wing with a flat top and curved bottom. This so-called "supercritical" wing—supercritical referring to that speed at which a large amount of drag is first encountered by an airplane traveling near Mach 1—delayed the formation of shock waves. The practical result of the supercritical wing's adoption was an increase in performance—improved fuel efficiency and greater range—rather than greater cruising speeds. The advance was quickly adopted by commercial airlines. (Although incorporation of supercritical wings can increase speed, nearly all commercial airlines have used the design to improve performance, thereby decreasing operating costs.)

In retrospect, the overriding aeronautical effort at Langley in the 1960s was research into a supersonic transport, or SST. After Langley's X-15 studies, it appeared as though an X-20—a so-called Dyna-Soar (for Dynamically Soaring Vehicle)—might be built to operate at speeds in excess of Mach 7 and that Langley would play a primary role in its development. But the Dyna-Soar project was cancelled in 1963. Fortunately for Langley's high-speed aeronautical researchers, by 1959, and as part of a joint NASA-Federal Aviation Administration effort, Langley had undertaken an SST technology-development program, known as the Supersonic Commercial Air Transport program, or SCAT. The aim of the SCAT studies was to identify ways a commercial supersonic transport could become part of the daily lives of American airplane passengers, as its subsonic sisters had.

The array of imperatives facing designers was intimidating. The SST would have to be structurally sound, fuel-efficient, cost-effective to operate, have a cruising speed of between Mach 2 and Mach 3, and not harm the environment. These difficult-to-meet and competing requirements were, ultimately, to prove too much for the then-current level of aeronautical technology to overcome, particularly in light of the ensuing political debate that sharply questioned the need for an American SST. In late May of 1971, the U.S. Congress cancelled the program, citing high cost of use, operational problems and environmental concerns.

Nevertheless, the effort brought together for the first time a number of space-age technologies: new metal alloys, new approaches

Richard Whitcomb looks over a model that incorporates his supercritical-wing concept.

Supersonic Commercial Air Transport (SCAT) model being readied for tests in the Unitary Plan Wind Tunnel.

to structural design, new engines, computer-controlled instrumentation, and computer-driven aircraft-design and environmental-impact modeling. In one sense, the SST program confirmed the modern dominance of the interdisciplinary approach in airplane design, a trend that has only intensified with the passage of time.

Five years after the American SST program was abolished, the British-French Concorde became the world's first viable commercial supersonic transport in regular service. An undeniable triumph of late 1960s engineering, the Mach 2 Concorde is still flying but has never turned a profit, limited as it is by passenger-carrying capacity, high operating cost and limited landing rights.

Shortly after the SST cancellation, Langley was directed to put its supersonic and hypersonic technology efforts into hibernation. That the Center kept the research alive (if barely) was tribute to the stubborn foresight that 20 years later would come in handy as the nation thought once again about propelling ordinary citizens faster than the speed of sound.

A SCAT model awaits aerodynamic evaluation.

Model of the supersonic transport (SST) variable-sweep version (with wings in the low-speed position) mounted prior to tests in the Full-Scale Tunnel.

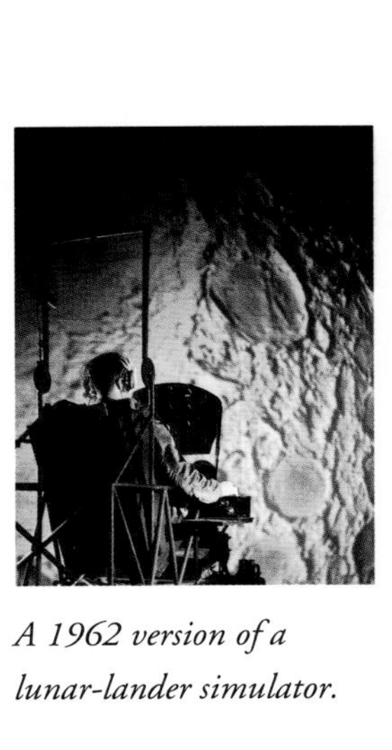

A 1962 version of a lunar-lander simulator.

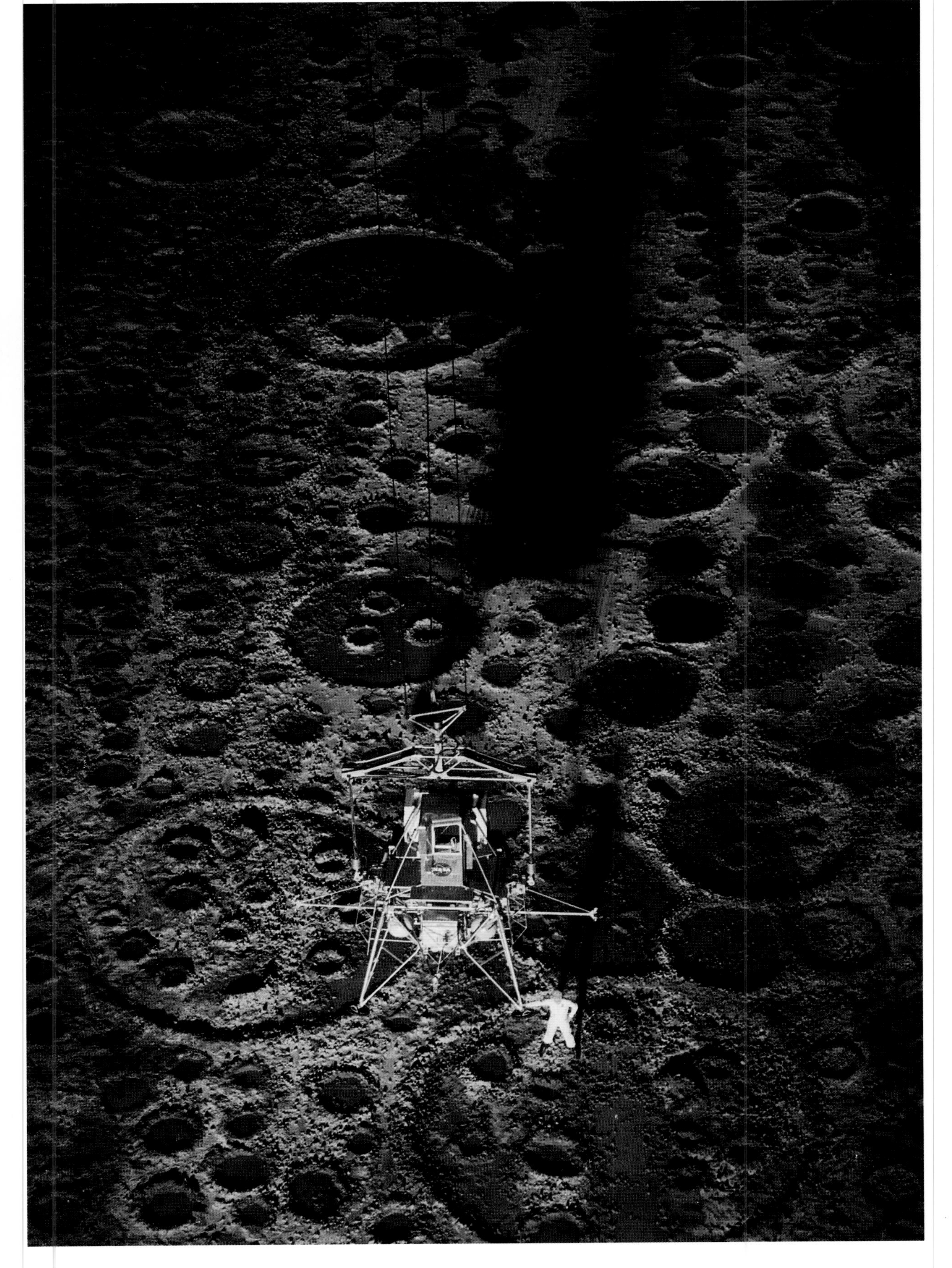

The simulated lunar surface of the Lunar Landing Research Facility, as seen from atop the facility.

Moon Matters

As Project Mercury began in the late 1950s, Langley was thrust full force into the national spotlight with the arrival in Hampton of the original seven astronauts. Under the tutelage of the Space Task Group, Scott Carpenter, Gordon Cooper, John Glenn, Virgil "Gus" Grissom, Walter Schirra, Alan Shepard and Donald "Deke" Slayton were trained to operate the space machines that would thrust them beyond the protective envelope of Earth's atmosphere.

The locals took keen note of Langley's astronaut-induced prominence. When Mercury proved successful, and ultimately evolved into Project Apollo, respect for the Center grew even greater, especially among the young. Adults, too, were caught up in the wave of enthusiasm. Hamptonians were so pleased with the attention that the space programs were bringing to their city that they voted to change the name of "Military Highway" to "Mercury Boulevard" and to dedicate the town's bridges in honor of the astronauts. Hampton and America had found new champions.

The Soviet Union, meantime, was moving forward determinedly with its space program. On April 12, 1961, cosmonaut Yuri Gagarin soared into a 108-minute orbit aboard the 5-ton Vostok rocket, thus officially becoming the first man to orbit the Earth. Three days later, the world's attention was refocused on Earth, as the American-led Bay of Pigs invasion of Cuba was repulsed by soldiers loyal to Fidel Castro. Following the fiasco, President John Kennedy sought to repair the damage done to the nation's prestige and his own political fortunes by intensifying America's space commitment. The result: the end-of-May 1961 speech during which the U.S. Moon mission was proclaimed.

Now that such an ambitious goal had been defined, the question was whether or not the United States could engineer its way to the Moon. Suborbital, even orbital, flights were doable. But by what method would a lunar

America's original seven astronauts trained at Langley in preparation for the Mercury Program. From left, front row: Virgil "Gus" Grissom, Scott Carpenter, Donald "Deke" Slayton and Gordon Cooper; back row: Alan Shepard, Walter Schirra and John Glenn.

Practicing with a full-scale model of the Gemini capsule in Langley's Rendezvous Docking Simulator.

landing be accomplished?

In order to meet President Kennedy's end-of-decade deadline, NASA considered three separate options. First studied was direct ascent, followed by Earth-orbit rendezvous (EOR) and, finally, lunar-orbit rendezvous (LOR). Direct Ascent involved the launch of a battleship-size rocket from Earth to the Moon and back again—basically the method popularized in Hollywood movies and science fiction novels.

A preparatory examination of the Lunar Orbiter spacecraft.

Practicing lunar-orbit rendezvous with the help of the Rendezvous Docking Simulator.

EOR entailed the launch into Earth orbit of two spacecraft, the payloads of which would be assembled into a vehicle that could travel to the Moon and then back to Earth.

The third choice was considered a dark horse candidate. According to the LOR concept, three small spacecraft—a command module, a service module (containing fuel cells, a control system and the main propulsion system) and a small lunar lander (also called the lunar excursion module, or LEM, this became formally named the Lunar Module, or LM)—would be boosted into Earth orbit on top of a three-stage rocket. Once in Earth orbit, the third stage of the rocket would then propel the craft's three-man crew into a lunar trajectory. Reaching lunar orbit, two of the crew members would don space suits, climb into the LEM, detach it from the mother ship and maneuver down to the lunar surface. The third crew member would remain in the command module, maintaining orbital vigil.

If all went well, after lunar exploration was concluded, the top half of the LEM would rocket back up to re-dock with the command module. After debarking from the craft, astronauts would then re-separate the lander's top half from the command module. The LEM would subsequently be cast adrift into deep space or deliberately crashed into the lunar surface to measure seismic disturbances. The three astronauts, safe and secure in the command module, would head for home.

LOR eventually prevailed over the direct ascent and EOR methods, mainly because of the efforts of a group of Langley researchers. In the opinion of many historians, LOR was chief among the reasons why the United States, in less than a decade, was able to manage humankind's first extraterrestrial excursion.

A rough approximation of spacecraft rendezvous in lunar orbit had been formulated as early as 1923 by German rocket pioneer Hermann Oberth. In 1959, Langley researcher William H. Michael, Jr., wrote an unpublished paper that briefly sketched the benefits of "parking" in lunar orbit the Earth-return propulsion portion of a spacecraft on a Moon-landing mission. Two separate groups of Langley researchers—the Lunar Mission Steering Group and the Rendezvous Committee—began to examine Moon-mission mechanics in 1959, using Michael's work as a point of departure. Working at first independently, and then together, the two groups became convinced that lunar-orbit rendezvous was NASA's best shot at lunar landing. NASA headquarters management, however, was not persuaded.

When Langley engineer and Rendezvous Committee head John C. Houbolt and a few of his colleagues initially approached NASA headquarters officials with the LOR idea, it was

Not long after this photo was taken in front of the Lunar Landing Research Facility, astronaut Neil Armstrong became the first human to step upon the surface of the Moon.

During a 1968 visit to Langley, then-CBS News anchorman Walter Cronkite tries out the Reduced Gravity Simulator, a series of cable-supported slings designed to approximate the Moon's gravity, one-sixth that of Earth's.

Launched from an overhead pendulum device, this one-fourth-scale model of the Apollo spacecraft was tested in the Impact Structures Facility to determine water-landing characteristics.

Lunar Orbiter II took this Moon shot of an area about as big as the combined states of Massachusetts, Connecticut and Rhode Island.

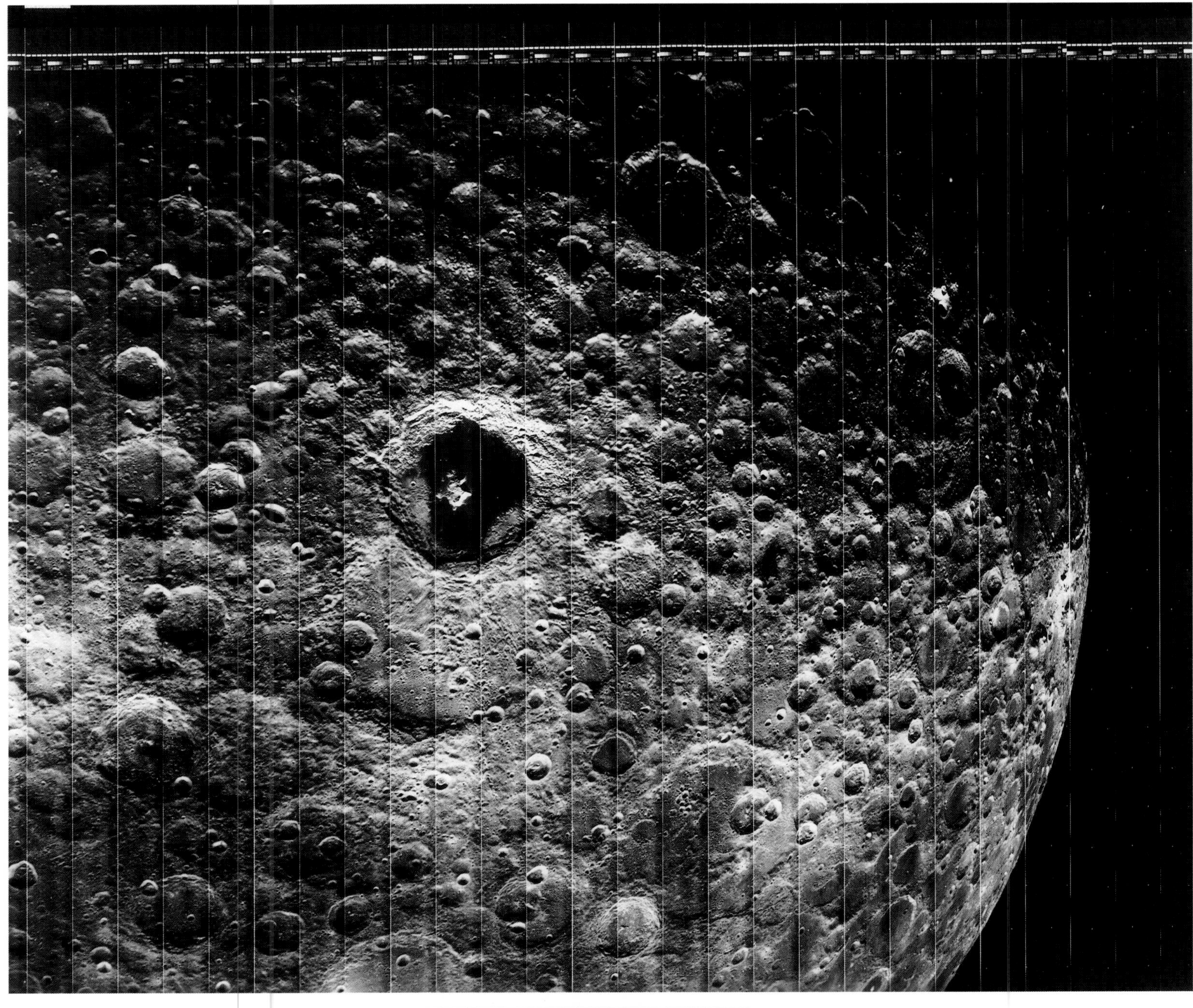

This photograph of the hidden, or "dark," side of the Moon was taken by Lunar Orbiter III during its mission to photograph potential lunar-landing sites for Apollo missions.

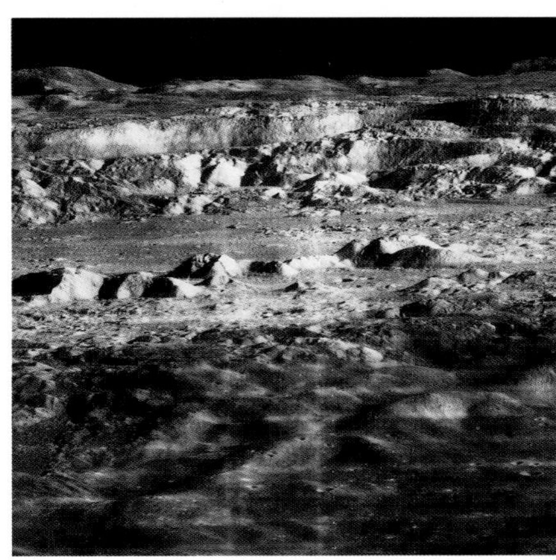

On November 23, 1967, Lunar Orbiter II's telephoto lens took this picture of the floor of the crater Copernicus. Copernicus, 60 miles wide and 2 miles deep, dominates the upper-left quadrant of the Moon as seen from Earth.

rejected as being unnecessarily complex and risky. Critics cited the danger: if the procedure should fail while the astronauts were orbiting the Moon, then they would forever be marooned in a metallic tomb. At least in the EOR scenario, if something went wrong, the astronauts could return home simply by allowing the orbit of their spacecraft to decay, reentering the atmosphere and then splashing down somewhere in an ocean.

Houbolt insisted and persisted, and after 2 years of sometimes heated discussions, NASA officials conceded his point: LOR was the way to go to the Moon. It would employ proven technology, incorporated a lighter payload, required only one Earth launch and would use less total-mission fuel than either of the other two methods put forth. Moreover, and importantly, only the small and lightweight LEM would have to land on the Moon. Part of LOR's appeal was also design flexibility; NASA could independently tailor all of the Apollo modules to suit mission requirements. In July of 1962 NASA administrator James Webb formally approved the LOR concept.

At a critical point in the early '60s, Langley researchers were the only ones in NASA fighting for LOR. It is difficult to say what the outcome might have been had the concept not been adopted. But the fact remains that, in less than a decade after President Kennedy's to-the-Moon directive, American astronauts were strolling the lunar surface.

Since practice makes perfect, there was a great deal of preparation for NASA's first "Moon shot." The Mercury program was the start. Astronaut Alan Shepard was the first American into space, although briefly; his suborbital mission lasted 15 minutes. John Glenn was the first American to orbit Earth, in February 1962,

In this 1967 photo, taken in Langley's 8-Foot High-Temperature Tunnel, preparations are being made to study reentry-heating effects on a nose cone design.

Also studied in Langley wind tunnels were the effects of wind and atmospheric turbulence on the Saturn rocket and escape tower, pictured here.

elating an American public eager for in-space success. After Mercury came Gemini, the project that would put to the test the maneuvers that would be required if Apollo was to be successful.

In particular, the Gemini astronauts would have to practice the rendezvous and docking techniques necessary to link two spacecraft. Accordingly, Langley built the Rendezvous Docking Simulator in 1963. Full-scale modules of the Gemini and Apollo spacecraft hung from an overhead carriage and cable-suspended gimbal system, the whole assembly being attached to the rafters of the Langley Center West Area Hangar. Astronauts "flew" the vehicles to rehearse and perfect docking skills.

Since the Moon is airless and its gravitational field is only one-sixth the strength of Earth's, there are no direct parallels between atmospheric flying and the piloting of a lunar lander. Some distinctly unusual problems would have to be overcome for the first manned lunar landing. For example, the thrust of rockets in the vacuum above the lunar surface would not produce the same effect as, say, that of rotating helicopter blades in air. Also, firing of control rockets could produce abrupt up-and-down, side-to-side, or rolling motions. The light would be different, too; the harsh glare of sunlight on the Moon's surface was unsoftened by an atmosphere, thereby throwing off depth perception.

To address these and other practical Moon matters, Langley built the Lunar Landing Research Facility (LLRF) in 1965. Twenty-four astronauts—including Neil Armstrong, the first human to walk on the Moon—practiced landings at this facility. Five-sixths of the weight of a full-size model LEM was supported by overhead cables, and thrust was provided by a working rocket engine. The LLRF base was modeled with fill dirt to resemble the Moon's surface and dark shadows were painted around the "craters." Floodlights were erected at the proper angle to simulate lunar light. A black screen was even installed at the far end of the gantry to mimic the airless lunar "sky." Neil Armstrong later said that when he saw his shadow fall upon the lunar dust, the sight was the same as he recalled while training at the LLRF at Langley. Attached to an overhead, lightweight trolley track that was part of the LLRF was the Reduced Gravity Simulator. There, suspended on one side by a network of slings and cables, an astronaut's ability to walk, run and perform the various tasks required during lunar-exploration activities was evaluated.

The Center built other equipment to imitate lunar conditions. A simulator constructed at the Center in the early 1960s helped researchers determine the ability of a pilot to control vertical braking maneuvers for landings, starting from an altitude of about 25 miles above the lunar surface. There was also a special facility that employed one-sixth-scale models of the lander to gauge the impact of landing loads. Another laboratory apparatus probed the anticipated and much feared problem of blowing dust caused by rocket blast, which could obscure the lunar surface and prevent the LEM pilot from locating a safe landing spot.

One of Langley's most noted achievements during this same period was the design and management of the Lunar Orbiter project. Third in a series of NASA-sponsored programs designed to choose the most suitable landing spot for Moon-landing missions, Lunar Orbiter photographed nearly all of the lunar surface in a series of spectacular close-ups. Some of the lunarscapes, of the far or "dark" side of the Moon, had never before been seen by the human eye.

On April 16, 1964, NASA signed a contract with prime contractor Boeing Corporation to construct Lunar Orbiter. Just 28 months later, on August 10, 1966, the first Orbiter blasted off on its ambitious trek. Eventually, five Lunar Orbiter spacecraft were launched. All five were successful. (The final launch occurred in August 1967.)

The craft essentially consisted of an 850-pound platform on which was mounted a built-to-order two-lens camera that took photos of the lunar surface on rolls of 70mm aerial film. The film was actually developed on board the vehicle, pressed into contact with a web that contained a single-solution processing chemical before it was "read out" and transmitted to Earth-based receiving stations. Ninety-nine percent of the Moon's surface was mapped by

The Apollo 11 Command and Service Modules are shown in a photo taken from the Lunar Module while in orbit around the Moon. The terrain below is the northeastern portion of the Sea of Fertility.

Lunar Orbiter. Of the eight sites identified by Lunar Orbiter III as appropriate, one—in the Sea of Tranquillity—was chosen as the place for the Apollo 11 landing.

At the time, Israel Taback was chief engineer and spacecraft manager for the program. He recalls an international conference in Prague, late in 1967, attended by astronomers from all over the world eager to see the photographic results of the Orbiter project. Taback was equally eager to oblige. Assisted by his wife, Taback unrolled large photo sheets of the lunar surface and covered them with transparent plastic. Then, on a gymnasium floor in a renovated 16th century school, and in their stocking feet, Taback and his colleagues went for a stroll on the Moon. "Sending off five spacecraft to orbit the Moon," Taback observes, "and then have them map the entire lunar surface ... well, it was an astounding thing at the time. And every one of them worked! It was thrilling."

The Center's space-race efforts also extended to wind-tunnel and general space-science research. Studied in Langley facilities were the effects of buffeting by wind, structural integrity, heat resistance and the durability of instrument design. Systems engineering personnel worked with other NASA centers on cooling, heating, pressure and waste-disposal systems. "We were working beyond the state of the art," says Barton Geer, retired Langley director for Systems Engineering and Operations. "Nobody had done things like this before."

Without Langley participation in the Mercury, Gemini and Apollo programs, there likely would have been no American Moon landing by mid-summer, 1969. As it was, on July 20 of that year, more than a billion people heard or watched Neil Armstrong take those first tentative steps upon another world. As he did so, Langley's entire staff could take justifiable pride in the indispensable role the Center played in a seminal event in human history.

"We had a target and a goal. Congress was behind it. Funding was available. The entire nation mobilized for a common goal," says John Houbolt, retired chief aeronautical scientist. "The landing on the Moon was undoubtedly mankind's greatest technological and engineering accomplishment. We started essentially from scratch in 1962 and 7 years later we were on the Moon. It was a remarkable achievement and remains unsurpassed."

On July 20, 1969, more than a billion people watched Neil Armstrong take humankind's first tentative steps upon another world.

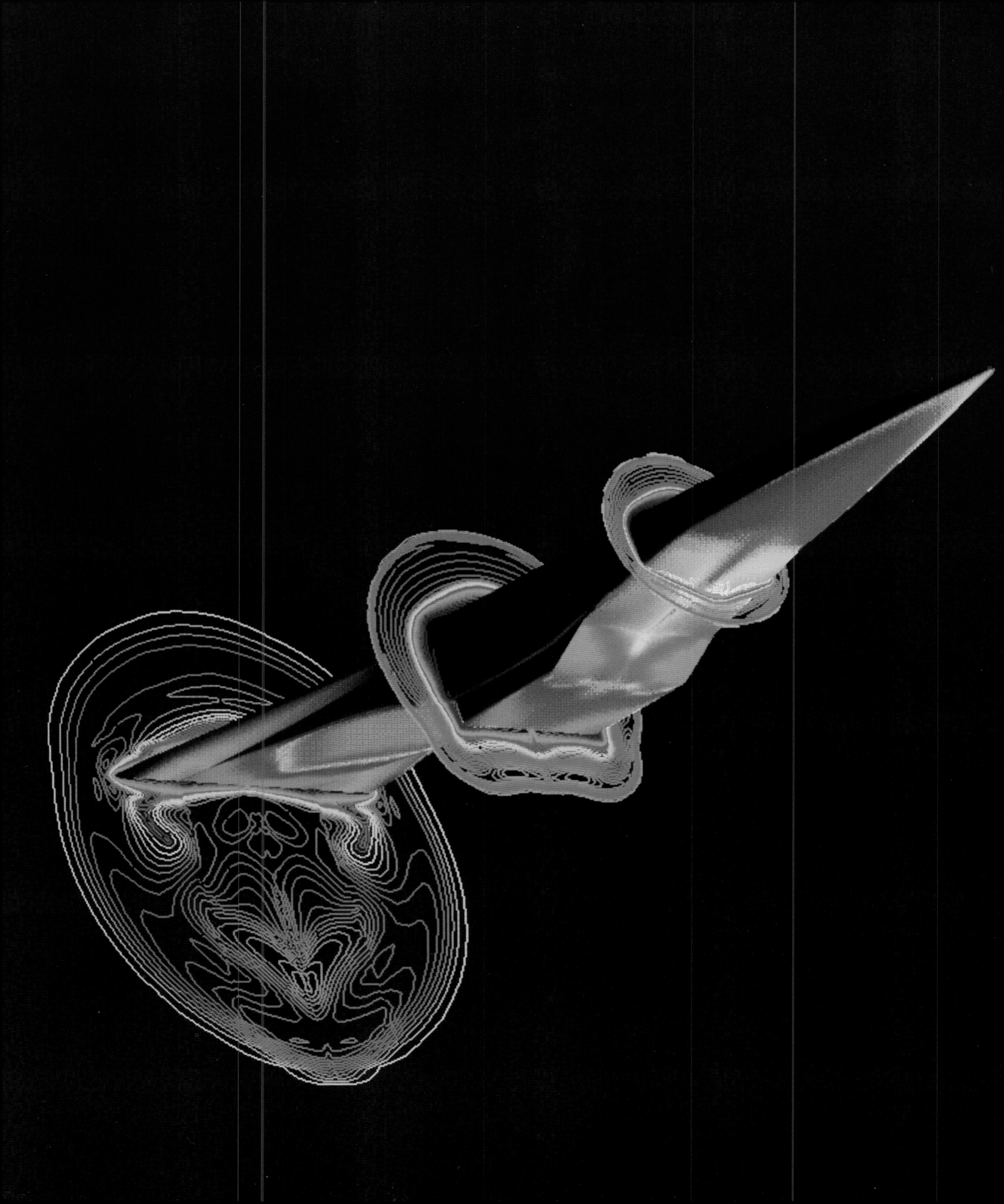

Charting New Courses 1970–1992 and Beyond

The most evocative images of the Apollo program are photographs not of the Moon, but of the Earth. Seen from a distance, Earth appears a startling oasis of life, a fragile bubble of animate color afloat in the ebony void of space. Apollo's revelation to the earthbound was of a home planet of great beauty, a world that, compared with the barren inhospitality of the rest of the Solar System, was a vivid reminder of the improbability of life.

In the aftermath of Apollo, it was sometimes hard to believe that a scant generation earlier interplanetary travel seemed the wildest fantasy. But by 1970, with men landed on the Moon and planetary probes beginning to open human eyes to otherworldly landscapes, perspectives were beginning to change. A larger, more exciting, more wondrous universe beckoned. What other marvels awaited humankind as it audaciously roamed beyond the planet of its birth?

A supercomputer-generated model of the airflow around one possible National Aero-Space Plane design traveling at 6,500 mph. Calculating the complex flows within a turbulent fluid, like air, is made possible by applying the mathematical rules of the Navier-Stokes equations, a process known as computational fluid dynamics.

Earthrise as seen from lunar orbit.

Even as humans took on the cosmos, there remained many vexing terrestrial problems. Those who lived through the period of the 1960s will recall conflict of all sorts: political, social, cultural, economic. The space program was not exempt from its share of controversy. Critics blasted Apollo as a flight of technological fancy that wasted precious dollars that otherwise could be spent bettering the lives of the disadvantaged. Supporters admitted that the space program was expensive but argued that the future payback, in terms of a deeper scientific understanding and improved technology, was enormous. The first part of that argument, of "spinoff" products from the space program benefitting the average citizen, was to be advanced more forcefully in coming years, as NASA was thrust into the relentless media glare and asked to justify every action and explain every shortcoming.

The Lunar Rover during the early part of the first Apollo 17 extravehicular activity.

As the Apollo program wound down, NASA seemed the victim of its very success. To use a time-worn sports analogy, the football game was over, the Super Bowl had been won: The Space Race finished, there was no longer any outer-space contest that needed winning. Some members of the legislative and executive branches of the U.S. government felt that since NASA had done the job President Kennedy required of it, the agency could now go back into its laboratories and finish whatever obscure research projects it

Launch of the Mars-mission Viking 2 payload on Titan III Centaur rocket, September 9, 1975.

wished—just as long as it didn't ask for a lot of money.

Nor, by the mid-1970s, did the American public seem all that interested in space anymore. After the wrenching national pain of Vietnam, an embargo imposed by oil producers in the Middle East and the arrival of "stagflation," it looked as though the United States might retreat from the space beachhead it had established. To be sure, there were impressive projects—Skylab, a joint U.S.-U.S.S.R. rendezvous linkup of the Apollo and Soyuz spacecraft, the development of the Space Shuttle—but few proposed any manned program on the huge scale of Mercury-Gemini-Apollo.

Meantime, at Langley, there was a period of belt tightening, of staff cuts and reduced budgets. In 1966, the Center employed some 4,300 civil servants, a figure that decreased by approximately 130 a year beginning in the early '70s; by 1980 the staff work force numbered 2,900. Large-scale projects were out and smaller, more focused programs with shorter term objectives were in. The role of contractors, made important during the Apollo years, increased. With less money to manage, Langley would have to establish priorities and decide how to balance the demands of aeronautical research with those of space science.

Characteristic of the more back-to-basics approach was an aeronautics program that began in 1972, when Langley joined with industry, university and U.S. Air Force representatives in an ongoing study of ways to incorporate so-called "composite" materials into new-aircraft design. Upon arrival of the 1973–74 energy crisis, this effort was redirected and renamed; the resultant Aircraft Energy Efficiency (ACEE) program sought to identify any and all ways to use airplane fuel more efficiently. The broad aim

Improving safety was one of the goals of crashworthiness tests conducted at Langley. Strapped-in crash dummies and a variety of monitoring devices installed in this general-aviation airplane provided crucial information to investigators.

The Space Shuttle scale models have spent more than 60,000 "occupancy hours" in Langley wind-tunnel tests.

was to provide an inventory of then-available and future technologies that could be used by aircraft manufacturers. The Center's ACEE research was more specific and concentrated in the areas of materials, structures and aerodynamics.

The U.S. manned space program was given a post-Apollo boost by the development of the first Space Shuttle, which underwent extensive developmental testing in the late '70s. With a long history of winged-vehicle experimentation—including research on so-called lifting bodies in the 1950s and 1960s—Langley took on primary design and aerodynamic research duties as the project went forward. In particular, Langley researchers were responsible for a crucial Shuttle design decision.

It was initially thought desirable to equip the Shuttle with jet engines that would drop into position as the craft reentered the Earth's atmosphere and maneuvered for landing. But Langley researchers argued in favor of a "dead-stick" landing, during which the Shuttle would glide, unpowered, to a runway touchdown. Center personnel pointed out that a dead-stick landing would be less complex, would reduce weight and would be safe besides.

(Researchers cited the experience of 300 pilots of Boeing jet transports, trained in dead-stick landings, all of whom validated the concept.) Although there was initial opposition to the Langley effort, NASA officials conceded the point as it became clear that the inclusion of jet engines would indeed increase the Shuttle's weight beyond acceptable limits. They were omitted from the craft's final design.

Langley also initiated a major Shuttle-support effort in its wind tunnels. There, Shuttle scale models spent more than 60,000 "occupancy hours" undergoing tests to verify aerodynamic soundness. Langley researchers conducted structures and materials tests, investigated and certified the craft's thermal protection system of glued-on tiles, developed simulations to solve problems in the Orbiter's

Aerodynamic testing of Space Shuttle and booster configuration at Langley's National Transonic Facility in 1985.

Space Shuttle Columbia *begins its fourth test flight in space on June 27, 1982.*

flight control and guidance systems, conducted landing tests on tires and brake systems and, later, participated in the redesign of solid-rocket booster components. Thus, when Space Shuttle *Columbia* soared to orbit on its April 12, 1981, maiden flight, Langley researchers could take considerable pride in the Center's contributions to the development of a new generation of spacecraft.

By the mid-1980s a new president, Ronald Reagan, had announced several aerospace initiatives, two of which—the building of a space station and the creation of the National Aero-Space Plane—would involve Langley Research Center directly. Reagan's successor, George Bush, had by 1990 outlined other ambitious plans, which included an increased American space-research presence in near-Earth orbit and a possible manned mission to Mars by the end of the second decade of the 21st century. These, too, would call upon the Center's research expertise.

Aerospace technology of the sort pioneered at Langley may better explain the workings of the home planet, even as humanity travels beyond it.

In the last decade of the 20th century a new chapter in astronautics was being written. Faster, safer, more environmentally benign aircraft were on the drawing boards, and there were plans to make spaceflight more economical and thus attractive to private interests. A spirited debate had been joined over how best to utilize terrestrial resources while protecting the Earth's biosphere; central to such discussions was how, for the first time in human history, to monitor and thereby understand planetary health. Aerospace technology of the sort pioneered at Langley seemed likely to be used not only as a means to better comprehend the workings of the home planet, but also to push beyond it, farther into space. While it didn't seem likely that the year 2000 would usher in the golden age predicted by some aerospace enthusiasts, there appeared little prospect of technological retreat from 75 year's worth of amazing aeronautical advances.

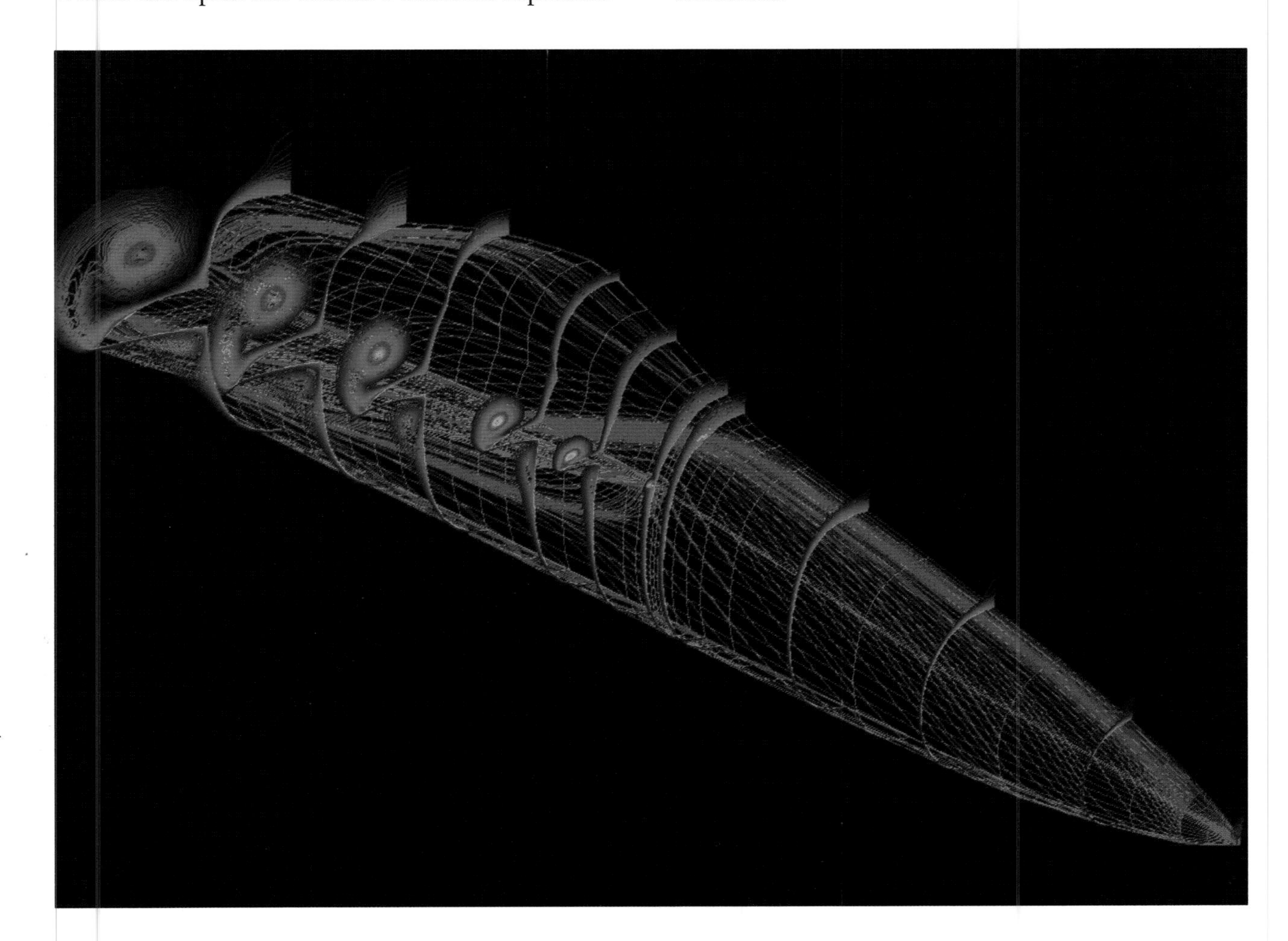

Supercomputer modeling of aerodynamic forces acting on a possible nose design for a hypersonic aircraft.

This boulder-strewn field reaches to the horizon, nearly 2 miles distant from Viking Lander 2's position on Mars' Utopian Plain.

Meeting Mars

In the 1870s and 1880s Italian astronomer Giovanni Virginio Schiaparelli identified features on the planet Mars that he believed to be an extensive system of canals. Schiaparelli and others theorized that "Martians" used the channels for irrigation, as aqueducts, or (like the Italian island city of Venice) for transportation. In later years the Martian "canals" were shown to be the result of poor stargazing equipment and fanciful imaginations. Still, and even into the last quarter of the 20th century, the question persisted. Was there intelligent life—indeed, any kind of life — on Mars?

That was one of the main questions the Langley-led Project Viking hoped to answer. Begun in the late 1960s, and the largest space-science undertaking at the Center since the manned space effort, Project Viking's goal was a soft landing on the surface of Mars followed by limited exploration.

The planet Mars as seen from Viking Orbiter 1 on June 18, 1976.

The ambitious project would confront engineering challenges not faced even by the complex Apollo program. Project Viking would entail the development of two different vehicles that would travel on one spacecraft. Once at Mars, and while both were still connected, the Viking Orbiter's job would be to select a landing site for the Viking Lander, conduct scientific investigations using the Orbiter's on-board radio system, and study the planet's topography and its atmosphere. The Lander's

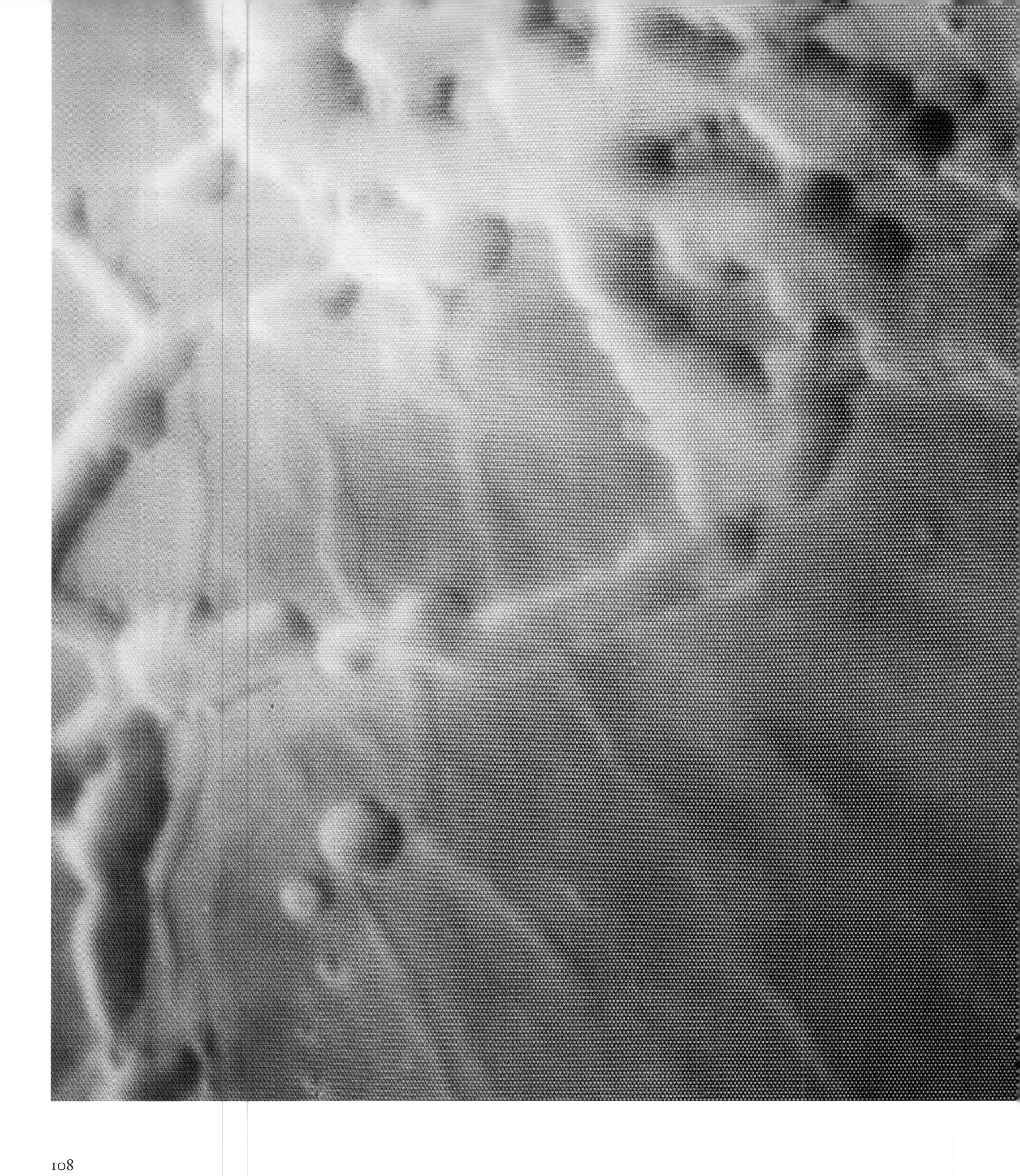

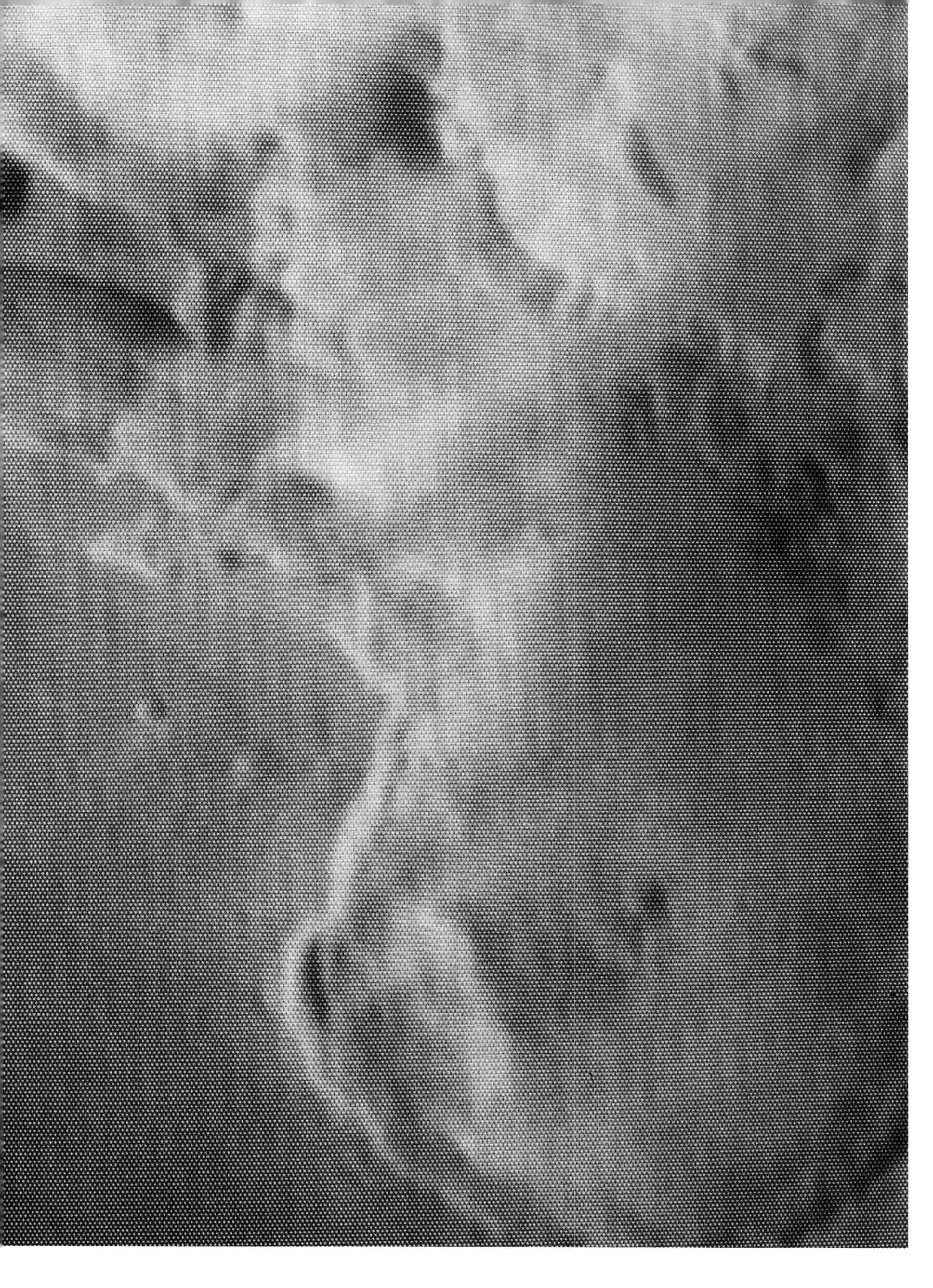

Taken during the Viking Orbiter 1's 40th revolution of Mars, this electronically transmitted image shows sunrise over the tributary canyons of a high plateau region. The white areas are bright clouds of water ice.

work was more demanding. Essentially a lightweight, rugged, automated extraterrestrial laboratory, it had to maneuver to a soft landing on the Martian surface and then undertake a series of studies on Martian geography, weather, chemistry and biology.

The Center asked for and received authorization to directly oversee the design and construction of the Viking Lander. In addition, Langley became the "lead center" for Project Viking. Langley coordinated the entirety of the work undertaken by other cooperating NASA centers, such as the Jet Propulsion Laboratory in California, which was itself overseeing the design of the Viking Orbiter. Langley was also given responsibility for construction management of the two vehicles and their constituent instruments, which were built by private contractors.

Closeup of the Martian moon Phobos taken by Viking Orbiter 1 on February 20, 1977.

Great technical sophistication was required to execute the scientific experiments, digitize the information collected, store the data, manipulate it, and then transmit it back to receiving stations on Earth. There was another, crucial requirement as well. "One of the most critical things was the sterilization of everything on the spacecraft," says Langley researcher Eugene Schult, who came to Langley in 1949 and who retired as deputy chief of the Center's Terminal Configured Vehicle Office in 1980. "That included all electrical components, every part of the structure, all the fluids. We had to sterilize to insure that Mars wouldn't be contaminated by any microbes imported from Earth."

Mr. Toad—part of the so-called Willows Formation—can be seen to the right of the large scoop the Viking Lander 1 has taken out of the Martian surface.

All of this complexity and sophistication had a direct dollar equation: developing such an intricate machine in such a small package against a specific deadline required a large budget. But the world in which NASA and Langley operated was full of budget restrictions. Even applying the lessons learned during one of Viking's Mars-probe predecessors, the Mariner program, and considering the dollar constraints, the task demanded enormous technological ingenuity and resourcefulness.

In the minds of a few Langley dissidents Viking was more of a research curse than blessing. Some of those on Langley's "aero" side were especially resentful of the resources sucked up by the project. "There were a lot of people in the research lab who hated Viking," confirms Paul Holloway, Langley director. "We were rebuilding aeronautics, taking on Viking and being hit by a gigantic manpower reduction, all at the same time. Viking had priority over everything and dominated all of our space technology efforts. There was a major impact on our research. Computers were tied up; wind-tunnel models couldn't get built. Yet even if it was one of Langley's most divisive projects, Viking was one of the Center's finest accomplishments."

Project Viking was not fated to answer all the questions posed by planetary scientists, but the fact that it addressed them "in person" was tribute to the engineering skill acquired at Langley after years of practice on such programs as Mercury, Lunar Orbiter, Gemini and Apollo. On July 20, 1976, on the seventh anniversary of the first lunar landing and 2 weeks after the 200th birthday of the United States, Viking Lander 1 touched down on the Martian surface. There, it and sister Lander 2—which landed on September 3—transmitted back to Earth spectacular images of the bleak Martian landscape.

In mid-August of 1976, less than a month after Viking Lander 1's Red Planet touchdown, the craft's sampler arm extended a retractable boom and pushed over the rock that Langley researchers had nicknamed "Mr. Badger." Labeled thus because of its shape, Mr. Badger was one of four rocks named in honor of the book *Wind In The Willows.* (Other rocks in the so-called Willows Formation were given the titles Mr. Rat, Mr. Mole and Mr. Toad.) Researchers were curious: would the soil under Mr. Badger be more moist than the surrounding, exposed soil? If so, perhaps there would be evidence of organic chemical processes, processes that could indicate the presence of primitive life. Unhappily for those hoping to find definitive proof of extraterrestrial existence, the outcome was not positive.

Designed to function for 90 days, all four Viking craft exceeded manyfold their intended operational lifetimes. Orbiter 2 was the first to fail, on July 24, 1978. Lander 2 ceased operation on April 12, 1980, followed 4 months later by Orbiter 1 on August 7. Lander 1 stayed "alive" 7 years past its design lifetime, until November 13, 1983, when it finally fell silent.

"To that day—maybe to this day—Viking was the most difficult unmanned space project ever undertaken. It brought Langley to the forefront of spacecraft technology," says Edgar M. Cortright, who arrived in Hampton in 1968 as Center director and who retired from that post in 1975. (Donald P. Hearth was appointed Cortright's successor and served as Center director until 1985.) "One of the most emotional experiences is to be part of a team that knocks itself out doing something worthwhile, and then succeeding. When Viking landed it was a real high. There was a tremendous mixed sense of exhilaration—that we did it—and relief—that it didn't fail. There was pride in the Langley team, pride in the accomplishment itself; it was the culmination of unbelievable effort. History was being made."

The first evidence of Martian frost—the white patches around the rocks—is revealed in this Viking Lander 2 photo taken on September 25, 1977.

Of Computers and Composites

Remembering that there was a time when American society didn't enjoy a widespread dependence on computers has grown difficult for a generation that relies on computational machines for the most ordinary of tasks, from banking to bill paying. Few realize that the idea of computing machines was introduced as early as the 17th century and that working models were on the drawing boards during the 19th. For a number of reasons, it was not until the 20th century that the first practical computers were introduced, in the 1930s. Another 40 or so years would pass before room-filling, power-hungry early models gave way to smaller, more compact designs. By 1974, and with the introduction of the "personal" computer, there began a momentum to miniaturization that, today, seems unstoppable.

To better aid their aeronautical investigations, Langley researchers rely on computers. The Center's Cray 2 supercomputer, pictured here, is capable of performing a half-billion calculations per second and allows researchers to conduct studies in three dimensions rather than two.

For researchers engaged in scientific inquiry, the advent of the computer has been a godsend, for it has revolutionized the way information is transmitted, stored and used. Modern computers enable the rapid, cross-connected flow of information so crucial to technological advance. Scientific exchange has accelerated to light speed, as researchers use personal computers, telephones, computer modems, electronic data bases, supercomputers, computer workstations, video equipment and facsimile (fax) machines in a constant quest to remain connected and in touch. What all of this may mean, in the long view of history, remains to be seen. In the short run, and at least in the field of aerospace, the impact has been great.

Full-motion simulators such as the one pictured here have added a realistic feel to the testing of new airplane control-and-instrumentation systems.

At Langley, computers have forever changed the way aerospace scientists and engineers do research. Langley researchers are using Center computers—which include, among others, Cray Y-MP and Cray 2 supercomputers and two "mini" supercomputers—both to create models of airflow around assorted aerodynamic shapes flying at varying speeds and to gauge an airplane's structural response to differing flight regimes. These studies in computational fluid dynamics (CFD) aim at predicting what will happen to a proposed aircraft design under real-world flight conditions. By evaluating variables long before a model is mounted in a wind-tunnel test section, computers have sped up the entire design-and-test process. "The whole design can be looked at to see how one change in one area affects all the others," says Frank Allario, the Center's director for Electronics. "In the past an aerodynamicist designed a particular shape. Then the structures people came in and built a structure around it. Then the controls people came in and fitted their instruments. Now the idea is to tackle the whole thing together, up front."

One of the biggest advantages afforded by computers is the real-time acquisition of data. Back at Langley's beginning, the engineers with the sharpest eyes would peer through tunnel observation ports, read the balance scales, and call out their readings to the individual acting as the recorder. It would be days, sometimes weeks, before the data were processed and the test results known. Using a computer—which can also be programmed to vary tunnel conditions, such as Mach number, air temperature and pressure—insures that researchers can be provided with up-to-the-second results of their investigations, thereby permitting ongoing adjustments to studies in progress.

At Langley, computers are also used to control the Center's simulators, which vary from full-motion devices to advanced versions of air traffic control systems. Langley's computer-controlled flight simulators create uncannily realistic in-flight conditions for pilots training on advanced fighter aircraft or for researchers testing modifications to flight systems, under controlled laboratory conditions and at much lower costs than actual flight testing. The Center's only in-flight simulator is the Advanced Transport Operating System (ATOPS), which is mounted in a full-size Boeing 737-100 and is used to make in-flight tests of next-generation aircraft-control configurations.

New cockpit displays make use of cathode-ray-tube (CRT) technology, the same method used to create images on TVs and computer screens.

Digital avionics are among the most ambitious aeronautical applications of computer power. Whether as part of flight control systems or navigation and guidance systems or employed to better orchestrate takeoffs and landings, digital avionics are changing the way airplanes fly. The inclusion of small television-like screens is one major advantage enjoyed on avionics-equipped aircraft; a wealth of easy-to-read information on flight conditions can be displayed thereupon for quick evaluation by pilots. "On-board computers can take real-time data and actually tell a pilot what can or should be done, especially if something goes wrong," says Jeremiah F. Creedon, Langley's director for Flight Systems. The rapid evolution of digital avionics is making for safer, easier-to-operate and, in the case of military craft, more maneuverable aircraft. From the early 1970s through the present, Langley has initiated or participated in a number of programs designed to evaluate these promising systems and their appropriate role in military and commercial aviation.

Test pilot Lee Person evaluates a "synthetic visibility system"—in essence, two helmet-mounted eyepieces connected to video cameras that swivel in response to head movements. The idea is to superimpose crucial information like airspeed, altitude and heading directly on the cameras' outside view.

Computers will also play a major role in the design of a 21st-century supersonic commercial transport, known as the High-Speed Civil Transport, or HSCT. Langley is one of the NASA centers participating in studies of the feasibility of such a craft. One major difference between the HSCT program now and the supersonic transport program of the '60s and early '70s is the Center's extensive use of its supercomputers in CFD modeling, allowing researchers to do in hours or weeks what would have previously taken months or years.

Fundamental to current HSCT research is the basic assumption that this new generation of airplane will make use of existing airports, fly on conventional jet fuel, meet allowable standards of airport noise, have no harmful effects on the atmosphere and be economically competitive with future long-haul subsonic airliners. Made from lightweight and

A model hypersonic craft undergoing tests in the 20-Inch Mach 6 Tunnel.

In this McDonnell Douglas conceptual design for a Mach 3.2 High-Speed Civil Transport, passenger capacity is approximately 300 and range is 7,485 statute miles.

very strong composite materials and exotic metal alloys, the HSCT would fly at speeds between Mach 2.5 and 3.2, cutting the travel time from Los Angeles to Tokyo from a current 10 hours to 4.

Another computer-intensive project at Langley concerns the National Aero-Space Plane, or NASP. Making a major splash when announced in 1985 by then-President Ronald Reagan, the mission for this hypersonic craft (initially misnamed the "Orient Express") would be to take off from conventional runways, travel to low-Earth orbit with freight and/or passengers, and then return for an airport landing. As conceived, the NASP would, in the latter stages of its orbit-insertion travel, attain speeds of Mach 25, or roughly 17,500 mph.

This metallic-looking corrugated shell is actually made from thermoplastic composite materials, graphite rods embedded in an epoxy matrix.

Langley is the lead NASA center for the NASP program and provides major technical support in several areas. Center researchers are evaluating practically every aspect of the proposed craft, from advanced materials to "scramjet" propulsion systems. In particular, Langley engineers are working on ways to integrate the NASP propulsion system with the craft's super-streamlined body.

If they are built, both the NASP and HSCT will be test beds for composite materials, the space-age substances that seem likely to supplement or even replace metals and metallic alloys by the middle of the next century. Called composites because they are composed of small rod-like fibers embedded in a binding substance, or matrix (similar to the way steel rods reinforce concrete), they promise a considerable improvement in performance, especially where airplanes are concerned. Since the early 1970s Langley has been in the forefront of composite materials research. Center researchers are seeking ways of employing composites in airplane structures and are working with aircraft manufacturers to identify the best means to do so.

"In 1970 there was no—not one—college course on composite materials at any university in the country," says Charles Blankenship, the Center's director for Structures. "Langley was the lead NASA center in getting these courses at universities. We've had to educate our engineers in a whole new field, in a new technology. And it's been quite an education over the past 20 years.

"In the past 40 years we've built a lot of things out of metal. We've come to know metal and its properties quite well. What composites offer us is more than one choice. Designers will have a lot of flexibility: They'll be able to use aluminum where it makes sense and composites where they make sense. There will be more options."

One of the most attractive features of composite materials is their weight-saving potential. Contemporary graphite-epoxy composites available from commercial sources demonstrate strength and stiffness as high as steel but at one-fourth the weight. Applied to full-body aircraft—wings, fuselage and control surfaces—the structural weight reduction could run as high as 25 percent, which would generate enormous savings in fuel costs alone. According to a NASA-commissioned study done in 1991, if current composite-materials technology were applied to the entire commercial U.S. aircraft fleet, the annual benefit would amount to some $2 billion.

Composites are quite resistant to structural fatigue—a small crack in a graphite-epoxy composite spreads much more slowly than one in aluminum, for example—and because they are nonmetallic, composites do not rust. The materials also have another major advantage: an ability to be precast into much larger, blended-body shapes, an example of which is a single part comprising wings joined to fuselage. This translates into a need for fewer fasteners and joiners, reducing parts cost and, in theory, permitting designers to routinely mass-produce at moderate price what today would be called custom-made airplanes. The use of composites is not yet widespread, at least in airplane manufacture. The price of the materials remains high, in large part because of labor-intensive manufacturing methods. Much also remains to be learned about the materials' durability over time and under adverse conditions. Langley is among those in the public and private sectors looking for ways to reduce composite-materials costs while validating real-world performance.

A Langley engineer checks the propulsion system inlets of a National Aero-Space Plane (NASP) model before testing begins in the 14- by 22-Foot Subsonic Tunnel.

Ultimately, future generations of aircraft may incorporate intelligent machine systems technology, also known as IMS. Such systems—computer-directed, built from composite materials and outfitted with sensors connected by fiber-optic "nerves"—would mimic the human body's own network of nerves and sense organs. Like humans, "smart" systems would be able to respond and adapt to a changing environment: to extremes of temperature and pressure, for example. One day, IMS-equipped devices may even be capable of limited self-repair. If such systems are ever built on a large scale—and Langley is testing small-scale IMS devices—then airplanes and spacecraft would undergo yet another remarkable design revolution.

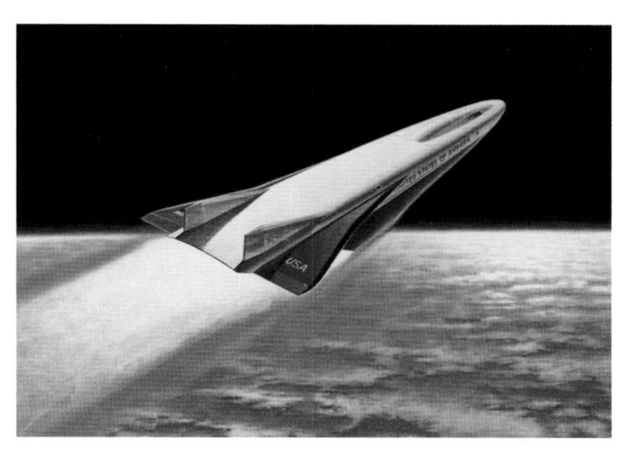

Artists concept of the X-30 aerospace plane flying through Earth's atmosphere on its way to low-Earth orbit. The experimental concept is part of the National Aero-Space Plane program. The X-30 is planned to demonstrate the technology for airbreathing space launch and hypersonic cruise vehicles.

The Science of Space and Air

Although both the United States and the Soviet Union have been orbiting either people or machines about the Earth since the late 1950s, there remains much to learn about the unique environment of space. How do materials and coatings react to near-constant bombardment by solar radiation or collision with extraterrestrial debris, like micrometeoroids? Do living systems fare well or poorly in almost total weightlessness? What are the effects of temperature extremes on organisms and structures?

Langley Research Center designed and built the Long Duration Exposure Facility, LDEF, to begin to answer such questions. Completed by 1978, tested for structural soundness in 1979, LDEF was shipped to the Kennedy Space Center in Florida in mid-1983 for a 1984 deployment by the Space Shuttle *Challenger*. The bus-size LDEF structure was outfitted with 57 experiments developed by more than 200 researchers, both in the United States and abroad. The investigators represented universities, private industry and government laboratories, including Langley and her sister NASA centers. Experiments fell into four broad categories: materials and structures, power and propulsion, science, and electronics and optics. LDEF would orbit Earth for 10 months as a "passive" satellite; those experiments needing power received it internally, from already affixed batteries or solar cells. No telemetry was transmitted to or received from the craft.

The explosion of the Shuttle *Challenger* in 1986 extended LDEF's mission life to nearly 6 years, as NASA reorganized Shuttle mission manifests in the aftermath of the tragedy. When LDEF was retrieved, in January 1990 by the Shuttle *Columbia*, the vehicle was seen as a virtual treasure trove by investigators eager to know how its cargo had weathered an inadvertently long orbital sojourn. A major preliminary finding revealed that outer-space structures made from composite materials will need a coating to protect them from micrometeoroids, space debris and degradation. Once on Earth, another LDEF experiment also bore fruit, so to speak: individuals and students worldwide were able to produce normal tomatoes from tomato seeds exposed to cosmic and solar radiation.

Far left: Deployed into orbit on April 7, 1984, by the Space Shuttle Challenger, *Langley's bus-size Long Duration Exposure Facility lived up to its name. Originally intended to stay in space 1 year, it was finally recovered in mid-January 1990 by the Shuttle* Columbia.

At the end of November 1985, an important Langley space-engineering experiment was put through its paces courtesy of the crew of the Space Shuttle *Atlantis*. Spacesuited astronauts, working from *Atlantis'* cargo bay, literally snapped together a 45-foot-long ACCESS (Assembly Concept for Construction of Erectable Space Structure) truss tower. The structure, which consisted primarily of tubular aluminum struts connected by joint-like nodes, was designed by Langley researchers and constructed by the Center's technicians. The purpose of the exercise was to determine the feasibility of future in-space construction techniques, tools and materials. The ACCESS experiment, concluded in about an hour, went smoothly and appeared to validate the practicality of in-space construction.

Tomato seeds are prepared for their launch aboard the Langley's Long Duration Exposure Facility (LDEF).

At Langley a variety of ongoing investigations aim to identify the best way to design, build and deploy large space structures, both manned and robotic. Some of those structures may house humans, like the proposed Space Station *Freedom*. Other constructs, like huge communications antennas, may be deployed to channel ever-increasing amounts of

This Langley-designed experiment, flown on the Space Shuttle Atlantis *in November 1985, was the first to demonstrate that a large trussed structure could be successfully assembled in orbit.*

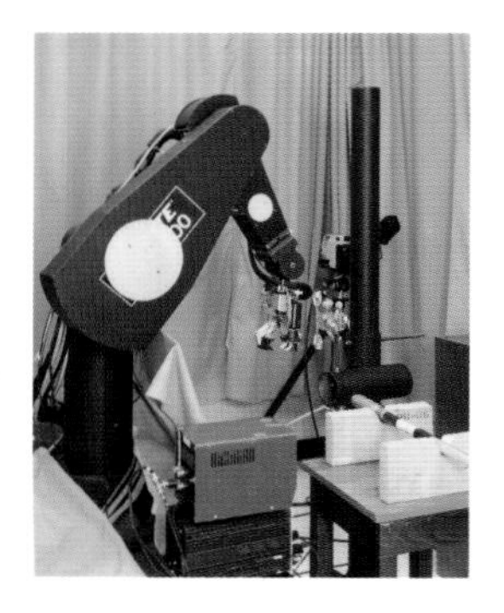

Evaluating robotic assembly as one future means of in-space construction.

The Earth Radiation Budget Experiment (ERBE) satellite pictured here was designed to measure and analyze fluctuations in the amount of heat energy emitted by the Sun and reflected or absorbed by the Earth.

data and information to distant points on the globe. Center researchers are in the process of developing automated systems that one day may assist human controllers in creating such high-frontier apparatuses.

Even as efforts continued in space, in the 1980s scientists were only beginning to understand the Earth's atmosphere and the complicated processes that maintain, renew and change it. Langley's Atmospheric Sciences Division (ASD) researchers are among those in the vanguard attempting to better comprehend the fundamental workings of the life-giving ocean of air that girds the planet. Formally organized in the 1970s, by the 1980s the Division had begun to examine the effect of clouds and cloud formation on global climate, the nature and extent of upper-atmosphere ozone depletion, the dispersion patterns and effects of trace gases (those that influence the so-called "greenhouse effect"), the atmospheric impact of large-scale burning of wood and vegetation (known as "biomass" burning), and the processes of global atmospheric chemistry in the Earth's lower atmosphere.

"We can take an idea, a glimmer in the mind," says Don Lawrence, chief of Langley's Atmospheric Sciences Division, "take it all the way through to building a device, flying or orbiting that device, and processing and then analyzing the data that results. At Langley we've built, I think it's fair to say, a world-class atmospheric sciences program."

Langley's atmospheric investigators have designed a wide array of sophisticated instrumentation, including customized combinations of lasers, telescopes and sensors that are flown on aircraft to measure extremely small concentrations of gases, small particles and water vapor. ASD scientists have fashioned satellite-based devices that gauge heat energy, have designed helicopter-borne instruments that analyze gaseous and solid emissions from fires, and are working on advanced sensing packages that will be orbited on future generations of satellites. Too, ASD researchers have devised software programs to analyze the enormous amount of data generated from ongoing global atmospheric experiments.

One principal ASD endeavor has been the design and management of the Earth Radiation Budget Experiment (ERBE), conceived in order to measure and analyze fluctuations in the amount of heat energy emitted by the Sun and reflected or absorbed by the Earth. Determining the whys and wherefores of the Earth's thermal equilibrium enable investigators to better understand the factors that drive world weather patterns and influence large-scale climate shifts.

The ERBE project was instituted in 1979, when LaRC's ASD scientists first began to outline the program's scientific objectives and devise the requirements for the instrumentation to accomplish them. From the outset, Division scientists managed the efforts of an international team of ERBE scientists and researchers, a team that is now some 60 members strong. One pivotal find was that clouds have a net cooling effect on global temperatures. "It really was a major scientific breakthrough," Don Lawrence says. "Now when climate modelers take clouds into account, they have quantifiable data to plug into their predictions."

In another major, ongoing effort, Langley's atmospheric scientists are examining how humans have modified the atmosphere that surrounds and nurtures life on Earth. Will man's destruction of vegetation and trees by burning have catastrophic consequences for this and succeeding generations? To answer such questions, teams of Center researchers have traveled all over the world to investigate the types and amounts of gases produced by man-made burning of grass, vegetation and trees. The emissions produced by such biomass burning are thought to add large amounts of carbon dioxide

Data from the Earth Radiation Budget Experiment (ERBE) has been used in this computer-generated image to indicate how cloud cover affects the amount of planetary heat radiated or retained.

and other greenhouse gases to the atmosphere.

Stratospheric ozone depletion has captured headlines, piqued the curiosity of average citizens and generated intense scientific effort. In 1985, a team of international scientists confirmed the existence of an ozone "hole" in a large region directly over the continent of Antarctica. Additional experiments have shown that ozone depletion is also occurring over the North Pole. Langley researchers have worked with colleagues all over the world to assist in plotting ozone-hole fluctuations. Indeed, a Langley study was the first to explain the mechanism by which ozone depletion is intensified.

Another of Langley's ASD-directed projects is the Halogen Occultation Experiment (HALOE), launched in mid-September 1991 on NASA's Upper Atmosphere Research Satellite (UARS). The UARS instruments are intended to measure concentrations of ozone, methane, water vapor, carbon dioxide, carbon monoxide, hydrogen fluoride and several types of chlorofluorocarbons (CFCs). HALOE is one of 10 separate instrument packages designed to provide atmospheric scientists with integrated global measurements of the chemistry, dynamics and energy flows throughout various regions of the atmosphere.

Atmospheric science is beginning to enter into its own. Practical spinoffs from basic scientific research, improved instrumentation, faster computers and a maturing space industry are fueling further research into the complicated functioning of the atmosphere and its interaction with Earth's vast oceans of water. Studies such as those conducted at, by and with Langley aim to identify, in unmistakably quantitative terms, the impact on the atmosphere of an ever-burgeoning human population. It is only through such studies that reliable information can be gathered, information that can be made available to citizens and policy-making boards for the tough public-policy decisions that will undoubtedly have to be made in the future.

A Langley researcher examines a readout of stratospheric ozone levels aboard a NASA research aircraft.

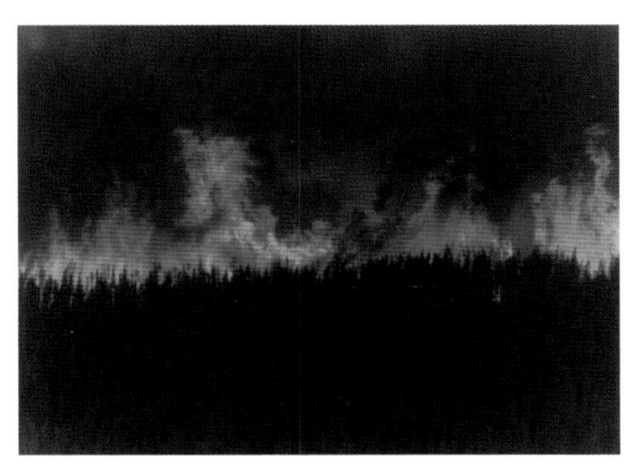

Langley atmospheric research scientists flew above this Canadian forest fire to obtain gas measurements relating to biomass burning.

The Breakthrough Business

When driving through heavy rain on interstate highways, few motorists today realize that their automobile travels have been made substantially safer by a research program undertaken at Langley Research Center. Begun in 1962, the Center's hydroplaning program (hydroplaning is the loss of traction on a water-covered surface) was originally intended to increase airplane tire traction, thereby decreasing braking distance. Langley's investigations concluded that the best way to help aircraft tires maintain firm contact on wet pavement was to cut thin grooves into that pavement, grooves through which excess water would drain. After tests in the late '60s and early '70s validated the concept, "safety" grooving was adopted for use on hundreds of airport runways around the world.

The practice also seemed appropriate for highways. Every state in America has since grooved at least part of its highway system. Too, safety grooves have been cut in pedestrian walkways, ramps and steps; food processing plants; work areas in refineries and factories; swimming-pool decks; and playgrounds. In 1990, the advance was selected for inauguration into the Space Technology Hall of Fame in Boulder, Colorado.

"It seems mundane when you think of it," says Cornelius Driver, who arrived at Langley in 1951 and retired in 1986 as chief of the Aeronautical Systems Division, "but grooving affects more people, and has saved more lives, than anything NASA has ever done. The amount of money put into [grooving research] was piddling, but the savings in human life and resources from highway grooving alone could probably pay for every one of the NASA budgets from day 1. In the broadest sense, such an accomplishment shows that government-sponsored research can have a tremendous payoff."

At the Aircraft Landing Dynamics Facility, a jet of water propels test carriages at aircraft landing speeds to measure stresses on aircraft landing gear and tires...

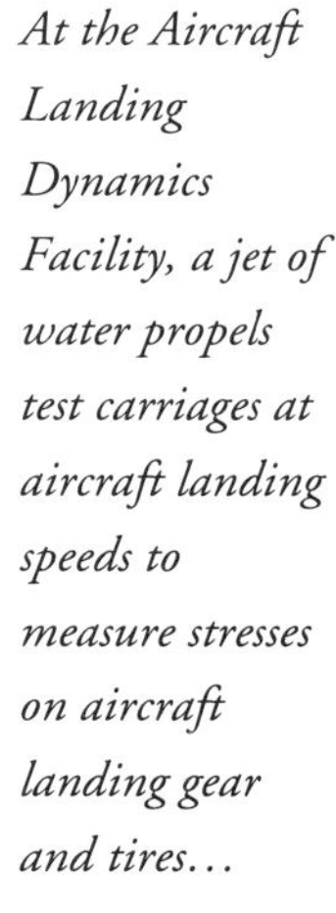

...and the result of such tests.

A closeup of grooves in the Wallops Island runway, carved to test their anti-hydroplaning effectiveness...

...and their installation on a California state highway.

When such programs as the one that resulted in safety grooving are described to Langley researchers as a "breakthrough," many are made uneasy by the word. They feel that it is too exaggerated a term to properly describe the Center's precise application of engineering science. The word "spinoff" is considered more appropriate in describing programs that result in innovative devices or procedures that have application in areas well beyond their original scope.

Take the case of the Center's "riblet" research. Building on marine-science studies into sharks' streamlined shapes, in the mid-'80s a Langley team found that V-shaped grooves a few thousandths of an inch deep reduced aerodynamic drag. That seemed promising enough, but the work caught the attention of yachtsman Dennis Conner, who was about to compete in the 1987 America's Cup. Conner eventually affixed to the hull of his craft *Stars & Stripes* a commercially produced thin plastic film grooved with thousands of riblets. In the words of the Australian skipper Ian Murray, whose yacht *Kookaburra III* Conner eventually defeated, the American thereby "found a tenth of a knot more than anyone else."

There have been other Langley spinoffs as well. Project FIRE work in the late 1950s and early 1960s led to the development of a furnace capable of melting metals for recycling. "Nondestructive" materials evaluation led to an ultrasonic device that uses sound waves to aid in the treatment of burn victims. Other notable examples of dozens of products that have been derived from work in Langley research facilities include a portable element analyzer that can detect such elements as gold, uranium, tungsten and copper; a hand-held plastic welding gun suitable for use in space; and a lightweight, composite-materials wheelchair for use on commercial airplanes.

Close examination of the skin of a fast-swimming shark appears to confirm Langley's aerodynamically efficient "riblet" concept. This view, magnified 30 times, reveals that projections on the shark skin—dermal denticles—line up to form grooves similar to those that have reduced drag in wind-tunnel tests.

Langley's man-made riblets.

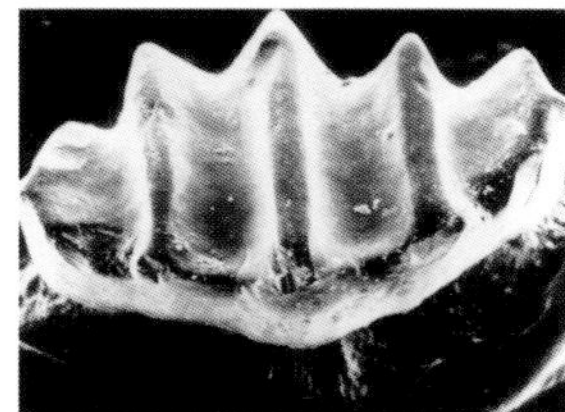

Mother Nature's riblet, this shark's dermal denticle has been magnified 3,000 times.

Research Spinoffs

Pick up a modern tennis racket, and you hold in your hand one of the many products to be born from aerospace research. The light weight and strength of the racket's graphite-epoxy frame owe much to the development and refinement of composite materials, space-age substances that are beginning to crop up in everything from airplanes to bicycles. Composites are among the multitude of products and processes "spun off" as a result of research sponsored by or conducted in NASA laboratories.

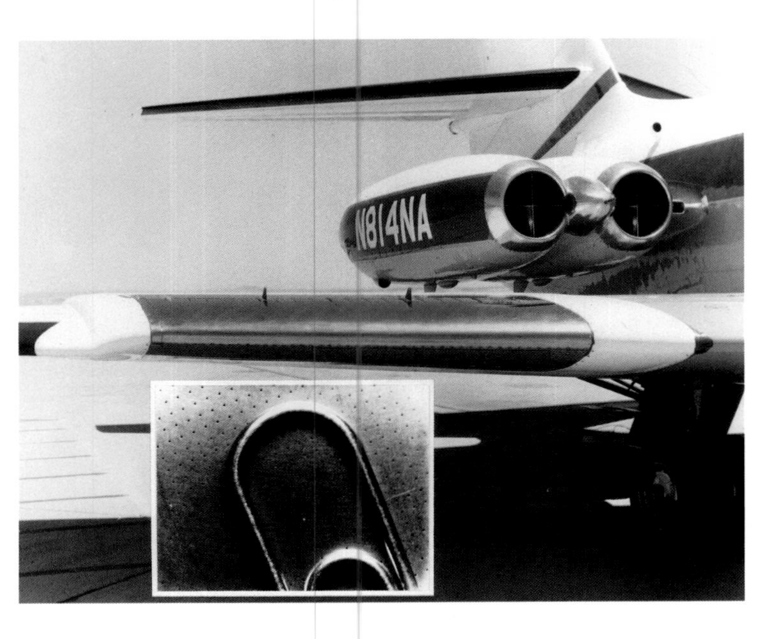

Tiny holes in the leading edge and surfaces of this airplane wing draw air in, creating a more laminar, or smoother, airflow, which in turn reduces drag and increases fuel efficiency. Inset shows a paper clip on the wing surface to indicate the relative size of the holes.

The primary agent of technology transfer from NASA's network of research facilities to everyday use is NASA's Technology Utilization Program (TUP), founded in 1962. Since TUP's founding, an estimated 20,000 to 30,000 spinoffs have found their way into the marketplace. A recent study conducted for NASA by the Chapman Research Group, Inc., examined in detail some 250 commercial uses of NASA-derived spinoffs—including automated blood pressure monitors, fetal monitoring devices, hang gliders, cordless power tools, pollution-control devices and high-temperature furnaces for materials recycling—and concluded that the total economic benefits have amounted to some $22 billion. But that figure may ultimately prove to be quite conservative. The spinoffs that were studied covered only an 8-year period, between 1978 and 1986, and represent a fraction of the 20,000-to-30,000 figure cited above. A follow-on study is currently underway to gather more comprehensive and accurate information.

Langley's Technology Utilization Office has been in existence since 1964. Since that time, Langley researchers have received numerous national and international awards for inventions derived from their work. Most Langley spinoffs have found both markets and buyers, and a number of former Center employees have gone on to found private companies to market spun-off products.

But spinoff seems too mild a word for the successful demonstration, in spring and summer of 1990, of a hybrid laminar-control system. Here, the word breakthrough might indeed be appropriate. In a joint project undertaken by researchers from Langley, the Air Force and Boeing Aircraft Corporation, a 22-foot wing section of a Boeing 757 was modified to test the effectiveness of "active" suction in reducing aerodynamic drag. Nineteen million small holes were drilled by laser into the 757's wing section, and a "Krueger flap" was added to the leading edge as an insect shield to keep the wing surface debris free during takeoffs and landings. Laminar, or smooth, air flow was achieved over 65 percent of the modified wing section—an unprecedented achievement in the down-and-dirty of normal aircraft operations.

If laminar flow were ever achieved over the majority of an airplane's surface, the fuel savings would be enormous; the drag caused by air friction on wings, fuselage, tail and engine nacelles could be reduced by at least 25 percent. Since estimates indicate that each percent of drag eliminated equates to an annual U.S. air-fleet savings of $100 million in 1990 dollars, $1 billion would be saved each year if all American commercial airlines managed a modest 10 percent reduction in drag.

The center section of each wing of this business jet has been modified for tests of laminar flow control.

Beyond Flight's Frontiers

Throughout its history, Langley has made a habit of going beyond the technologically expected. Most of the Center's work has not been of the breakthrough variety, in the common usage of the word. Nevertheless, over three-quarters of a century, the Center's methodical precision brought about great and beneficial changes both to airplanes and spacecraft. Langley led or was a major participant in aerospace innovation amidst an astonishing century-long explosion of science and technology.

In a Langley-directed study, an F-106 aircraft flies through storm clouds to measure the effects of lightning on electronic controls.

The progression of flight from the spray-drenched sands of a cold North Carolina beach into an even colder interplanetary void was epic and almost unbelievable. After all, in less than three generations, the drone of wooden propellers had been drowned out in the roar of jets and in the Earth-shaking fire of rockets. Never before in human history had such a long technological leap been made in so short a time. Throughout, as a place of aeronautical engineering excellence, Langley, as James Hansen writes in *Engineer In Charge*, "made change into a habit, and the expectation of surprise into a rule of thumb."

As one century ends and another begins, it is becoming more difficult to separate the word "aero" from "space." At Langley and elsewhere a new generation of aerospace engineers is beginning to consider the types of craft that, in coming decades, will breathe and fly through air and ply the vacuum of space. Like their predecessors, this generation of Langley engineers and theoreticians will be confronted by seemingly intractable difficulties. How they resolve them is for future historians to evaluate. Perhaps they will be regarded as the next wave of problem-solvers, the ones who figure out how to beat the gravity-well problem—the difficulty of inexpensively boosting payloads and people into orbit—once and for all. Or perhaps not.

For their part, the old guard remain skeptical. They ask whether young researchers who display an almost fanatical devotion to computers really understand what it takes to make an airplane fly better or improve the performance of spacecraft. There is also way too much bureaucracy in government service, say the old-timers: too much red tape, too many NASA managers chasing too few projects. What has been lost, these veterans grumble, is the hands-on, technical savvy of NACA times, when the engineer in charge himself wasn't afraid to go into a wind tunnel to get the job done.

This supersonic combustion ramjet, or scramjet, engine is put through its paces at four to seven times the speed of sound in Langley's Scramjet Test Facility.

Times, though, have changed. The issues confronting those on the cutting edge of aerospace research are more resistant to short-term resolution. Improving the speed, structure and handling characteristics of a fabric-covered biplane was accomplished in a relatively short period of time, but devising practical, economical designs for working supersonic and hypersonic airplanes is far more difficult and time-consuming. But just as Langley-led studies resulted in much-improved subsonic aircraft, so in time will the Center's investigations probably lead to commercial aircraft that will travel at several times the speed of sound.

Whatever difficulties await them, Langley's people appear to have retained the basic sense of excitement that exploration encourages. "Here our people know about, or are involved in, everything that's going on in aeronautics in this country," says Langley Director for Aeronautics Roy V. Harris, Jr. "We have unique access. Our people have the freedom to develop their technological intellect in a way that hardly exists anywhere else. That builds a sense of excitement. If it's aeronautics and research you're interested in, Langley's the place."

A one-eleventh-scale model of a Boeing 737 is prepared for testing in Langley's Low Frequency Antenna Test Facility as part of an effort to develop a collision avoidance system for commercial airliners.

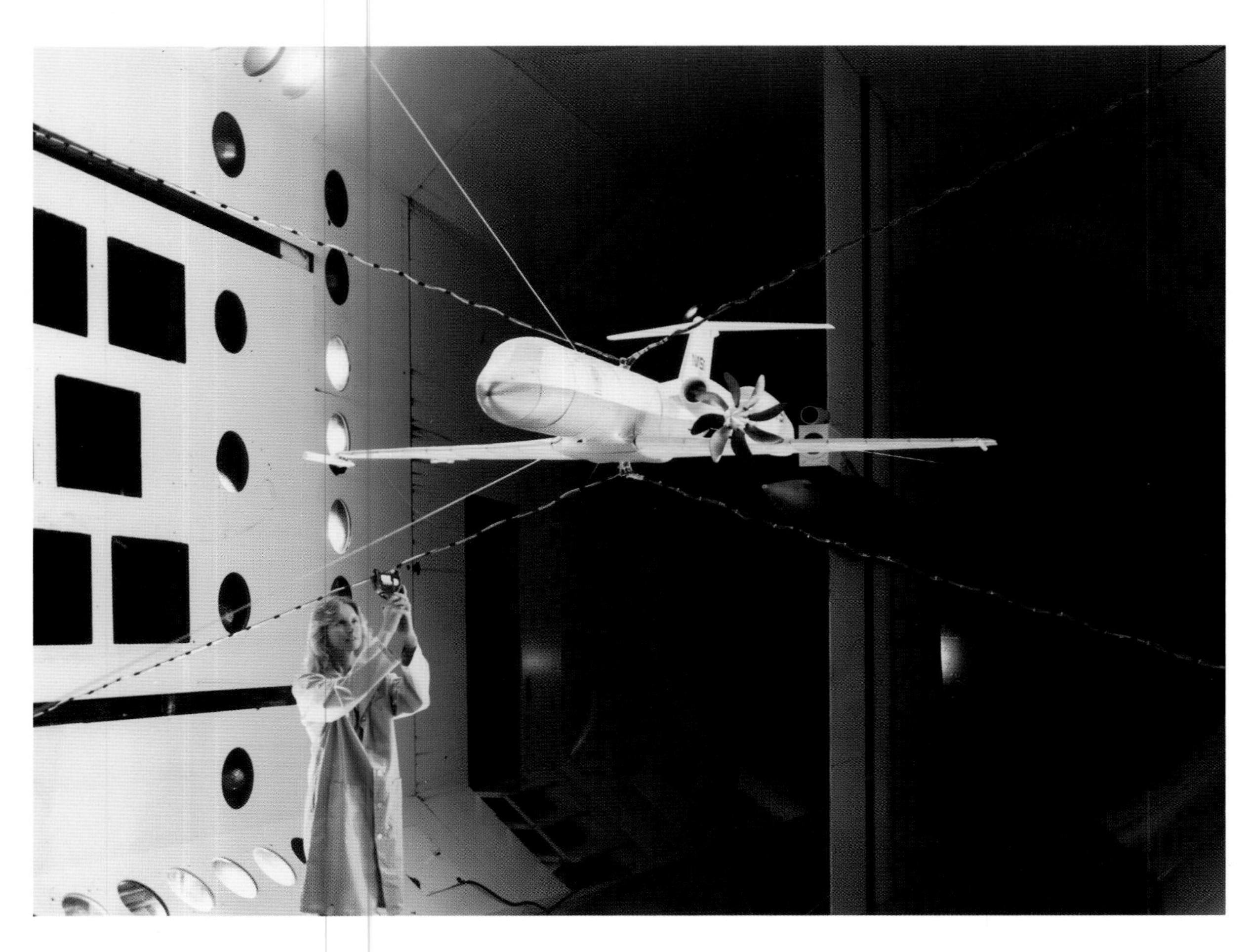

A one-ninth-scale model of a Gulfstream business jet with advanced turboprop engine (on left wing) undergoing flutter tests in the Transonic Dynamics Tunnel.

haven't been over *if* we could do it; they were over *how* to do it. The motto is, if you draw a picture of it we can build it. When a problem comes up, we can put together a team to work on virtually anything."

That work goes on. Center researchers are developing devices that will appear on, and techniques that will be incorporated into the design of, aircraft and spacecraft of the 21st century. What exact form those future machines will take remains unclear. In the main, predicting the march of science and technology has proved to be a notoriously dicey proposition. What does seem assured is that the pace of technology, already brisk, will quicken in the years ahead.

Stroll through the halls of Center buildings, peek into its workspaces, listen in the cafeteria: one is given the impression that the good ideas are still bubbling to the surface in Langley's daily give-and-take. It's an atmosphere not so different from the 1930s and 1940s, when engineers used to punctuate their animated discussions by scribbling equations on marble-topped lunchroom tables. At Langley one still gets the impression that, given time, the proper attitude and logistical support, most things are possible.

"At Langley there still is a can-do attitude," asserts Center Director Paul Holloway. "Over the years the arguments

Advanced-concepts model plane with front canards, winglets and pusher propellers, in 12-Foot Low-Speed Tunnel in 1984.

Anticipated for launch sometime before the turn of the next century are satellites that will comprise the ambitious Earth Observing System (EOS), part of the program known as Mission to Planet Earth. Designed to monitor the atmosphere on a continuing basis, EOS sensors will send back a steady, comprehensive stream of information about atmospheric workings. Atmospheric Sciences Division researchers and their systems-engineering and electronics-research colleagues at

Langley will play a prominent role in the development of complex EOS instrumentation and are already at work on the first phase of the project.

Other ventures seem ambitious indeed: orbital construction of space habitations and vehicles, a permanent lunar base, a manned mission to Mars. If these projects materialize on even half the scale envisioned, they will advance considerably America's technological prowess on the high frontier. Langley's exact role in these ventures is yet to be determined, but given the Center's past and present engineering expertise, it is likely to be fundamental.

One day Langley may well be engaged in research relating to a manned mission to Mars or helping to design craft that will conduct scientific research from manned outposts on the satellites of Jupiter and Saturn. Or perhaps the Center will concentrate on projects closer to home, figuring ways to fly ever faster and more safely through Earth's atmosphere and designing the next generation of automated, unmanned space probes. No matter where aerospace research ends up, it seems certain that Langley will continue to do what it has done best: figure out what works, and works better, and then make sure the improvements find their way in due course onto the machines that fly in the air and travel through space.

One wonders what the Wright brothers would have made of supersonic transports, of supercomputers, of Moon shots and planetary flybys. One hopes the enterprising pair would have approved, if not of the complexity or cost, then perhaps of the spirit of adventure and the thirst for knowledge such endeavor provokes. From Langley Research Center's perspective, tomorrow cannot be clearly seen. What is sure is that tomorrow's challenge, and all the frustration and fulfillment it will bring, is an inevitable fact of life. For those working beyond the frontiers of flight, that is reason enough for celebration.

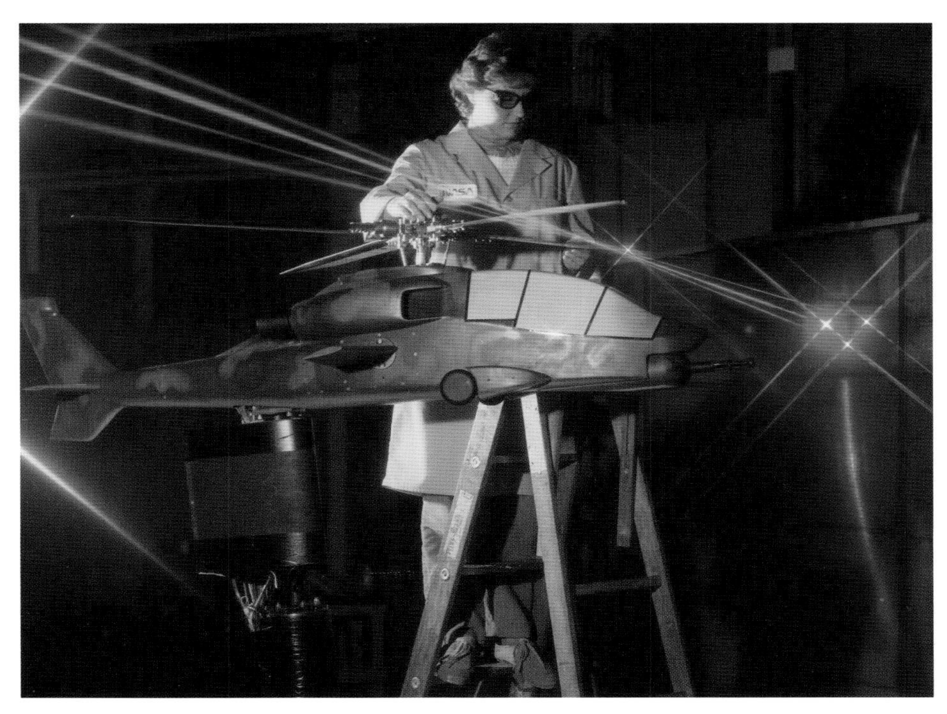

A researcher aligns an advanced helicopter model prior to laser-assisted tests in the 14- by 22-Foot Subsonic Tunnel. Investigators make use of a laser velocimeter, a device that measures complex airflows, thereby helping to predict helicopter rotor performance.

Closeup of the nozzles that inject nitrogen gas into the National Transonic Facility.

Aeronautical Breakthroughs in a Century of Flight

Army Curtiss AT-5A was first plane fitted with NACA cowling.

America's first jet airplane, the Bell P-59.

1928 **1942** **1944** **1947**

Republic P-47 Thunderbolt.

The Bell XS-1 broke the sound barrier on October 14, 1947.

The HL-20 experimental aircraft mock-up.

The X-15 experimental aircraft.

1959 **1963** **1991** **1992 and beyond**

Navy combat air patrol wind tunnel model shows two extreme positions of variable-sweep wing.

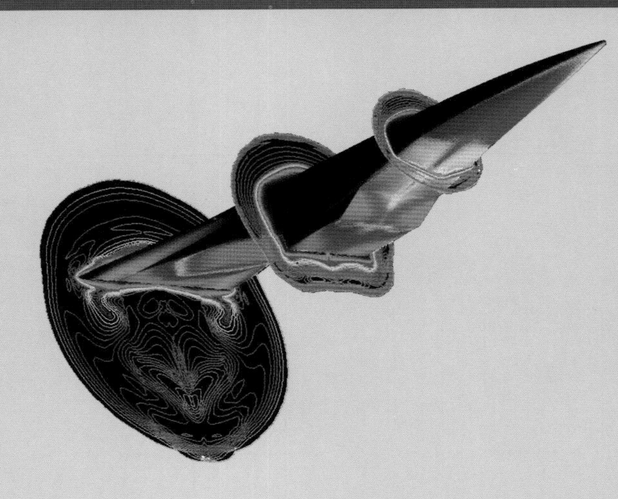

SCRAMJET engine exhaust is modeled in this supercomputer-generated image of an aerospace vehicle as part of the National Aero-Space Plane (NASP) Program.

Suggested Further Reading

Baals, Donald D. and Corliss, William R. *Wind Tunnels of NASA.* U.S. Government Printing Office: NASA SP-440, 1981.

Becker, John. *The High-Speed Frontier: Case Histories of Four NACA Programs, 1920–1950.* U.S. Government Printing Office: NASA SP-445, 1980.

Compton, William David. *Where No Man Has Gone Before: A History of Apollo Lunar Exploration Missions.* U.S. Government Printing Office: NASA SP-4214, 1989.

Ezell, Edward Clinton and Ezell, Linda Neuman. *On Mars: Exploration of the Red Planet, 1958–1978.* U.S. Government Printing Office: NASA SP-4212, 1984.

Hansen, James R. *Engineer In Charge: A History of the Langley Aeronautical Laboratory, 1917–1958.* U.S. Government Printing Office: NASA SP-4305, 1987.

Loftin, Laurence K., Jr. *Quest for Performance: The Evolution of Modern Aircraft.* U.S. Government Printing Office: NASA SP-468, 1985.

Nicks, Oran W. *Far Travelers: The Exploring Machines.* U.S. Government Printing Office: NASA SP-480, 1985.

Roland, Alex. *Model Research: The National Advisory Committee for Aeronautics, 1915–1958.* U.S. Government Printing Office: NASA SP-4103, 1985.

Smithsonian Institution: National Air and Space Museum. *Milestones of Aviation.* Hugh Lauter Levin Associates, New York, N.Y., 1989.

Swenson, Loyd S., Jr., with Grimwood, James M. and Alexander, Charles C. *This New Ocean: A History of Project Mercury.* U.S. Government Printing Office: NASA SP-4201, 1966.

Credits

Dozens and dozens of people played a role in preparing, researching, verifying, and overseeing the *Winds of Change.* A "thank you" goes to each of these unsung heroes, with special recognition to the following contributors whose efforts were vital to the success of the book.

Many technical experts, all NASA Langley Research Center retirees, shared endless hours reviewing and providing information for the book that added to its technical validity and general historical significance. These people were Donald D. Baals, John V. Becker, John E. Duberg, John C. Houbolt, Samuel Katzoff, Edwin C. Kilgore, Laurence K. Loftin, Jr., Eugene Lundquist, Axel T. Mattson, William H. Phillips, John P. Reeder, Israel Taback, Richard T. Whitcomb, and Charles H. Zimmerman.

Special consultants, who reviewed the book and provided valuable input, were Edgar M. Cortright and Donald P. Hearth, both former Directors of the Langley Research Center, and James R. Hansen, Langley historian.

Current NASA Langley researchers helped keep a balance between past achievements and Langley's current work. These men were Gary P. Beasley, James E. Bostic, Reginald M. Holloway, Robert J. Huston, John K. Molloy, Edwin J. Prior, and Willard R. Weaver.

Still other NASA professionals contributed to the book in a variety of ways. Howard S. Golden and Robert Schulman, from NASA Headquarters, contributed immensely to the overall production and design of the book. Langley's Thomas H. Brinkley and Mary K. McCaskill provided technical editing support; Elizabeth G. Fedors contributed to graphics-related efforts; Richard T. Layman reviewed and supported historical requirements; and A. Gary Price was the on-site consultant. Project oversight was provided by Karen R. Credeur, advisory committee chairman, and Catharine G. Schauer, the project manager.

Others, non-NASA folks, shared their expertise to help make the *Winds of Change* representative of Langley's 75 years. Ralph T. Johnston, Director of the Virginia Air and Space Center, provided comments both as an aeronautical buff and as a museum expert; and Stephen E. Chambers and Lynn Van der Veer, who formed the art and graphics design team that transformed words and pictures into *Winds of Change.*

Finally, "thanks" are given to: Paul (Mike) Willis, NASA printing specialist, for handling the complex production paperwork; the Government Printing Office's Technical Review Section for quality controlling *Winds of Change* through proofs and press; and to Peake Printers, Inc. for an outstanding example of the lithographer's art.